I have tried to recreate events, locations, and evidence to the best of my recollection. Events with parties mentioned were witnessed firsthand. In some instances I have chosen to maintain their anonymity to protect their privacy. Although the author and publisher have made every effort to ensure that the information in this book was correct at press time, the author and publisher do not assume and hereby disclaim any liability to any party for any loss, damage, inaccuracy, or disruption caused by errors or omissions, whether such errors or omissions result from negligence, accident, or any other cause. Opinions and theories in this book are the opinions of the author and do not necessarily reflect the opinions or ethics of any parties mentioned.

Original cover art design by Jean-Francois Podevin @ podevin.com

WHEN THE DEAD SPEAK

The Art & Science of Paranormal Investigation

Kitty Janusz

"The boundaries which divide Life from Death are at best shadowy and vague. Who shall say where the one ends, and the other begins?"

Edgar Allan Poe

(1809-1849)

Dedication

This book is dedicated to my good friend and mentor, Kimberly Rinaldi, who inspires (and pushes me) and in loving memory of my good friend Debby. You taught me so much about capturing EVPs and investigating the paranormal respectfully. I miss you every day..

Table of Contents

Introduction

Mankind has always asked the big question: What happens when we die? Are we just food for worms at that point? Early Man must have believed there was some sort of self or consciousness that remained after the body was gone.

Early civilizations documented great respect and care be given the body after death. They felt it important to carry out elaborate burial rituals and ceremonies. The deceased were buried with the trappings of life, to carry them over on their journey. Food, jewels, pets, sometimes even servants were buried alongside a person of high stature so they would be comfortable in the afterlife.

So where DO we go? Do we meld into the Great Collective, leaving our soul and personality to simply dissipate? Can we remain on Earth in spirit form, retaining our memories and personality? Can we decide if we want to remain on this plane and cause cold spots and wreak havoc on people who drove us crazy when we were alive?

If you've ever admitted to someone that yous're into the paranormal, and have had them respond with, "There's no such thing as ghosts," what did you say in response? Did you ask if they had spent any time studying parapsychology and quantum physics before coming to this conclusion? My guess is that if you did, the answer would be no.

Well, my response to this statement usually snaps them back to reality: "Well, how exactly do you define "ghost'?"

Making someone just think about the definition of a ghost or spirit gives them pause. It's impossible to deny the existence of something you can't even define. So how would someone like me, who has photographed them, felt them, seen them, and conversed with them, define ghost? Let's start by saying I believe that all forms of life are made of energy pulsating at particular vibrations. The higher and faster the vibration, the higher the ascension and closer to perfection. The higher vibrational beings are closer to creation, be it God, the

Great Collective, or whatever you consider the Divine Source.

Going back to ghosts, I measure entities by a rising scale. Lowest on the rung are demons. Demons never walked the earth in human form. They embody hatred and negativity. They drain energy from other sources which in humans can cause sickness, mental illness, depression leading to oppression, and full possession. In extreme cases, they can kill. I choose not to engage with anything that I feel may be demonic. I don't want to give it the satisfaction or attention. If someone needs an exorcism, your neighborhood Catholic church should have a contact for a demonologist. In fact, I think it's a requirement for every Catholic church to have a demonologist on retainer, just in case.

Next on the ladder are elementals. Elementals are entities of limited consciousness. They also never walked the earth in human form. Elves, sprites, fairies, and wood nymphs all fall into this category. I'm not saying these are all bad. I know folks who love working with fairy energy and leave Skittles in their yard to entice them. But some elementals are ancient earth energy. Elementals can be very troublesome if given free reign. They quickly develop a sense of entitlement to a person or location and can wreak havoc. Often seen as small, black masses, elementals enjoy chasing household pets and rearranging your pots and pans. Think Gremlins, not Tinkerbell.

Next up are what I call a true ghost. A true ghost is an earthbound spirit. A ghost walked here in human form but has not crossed into Light or been reincarnated. Either they are unwilling to move on, feel undeserving to move on, or have no idea what's going on. They are stuck. There's a school of thought that says earthbound ghosts stay because they have an agenda. Perhaps there is some unfinished business before they can be at peace, but sometimes not. Some believe this group makes up most of the pranksters and bullies in the afterlife. They can be manipulative and predatory. Bully ghosts do exist, subduing weaker spirits and picking on vulnerable living folks. When here in the physical world they may have been thugs and bullies and they don't always change their personalities after death. They can screw around with you and your equipment. They can affect your health to some degree, making you nauseous or irritable. Ghosts may scratch you or even push you. Ghosts are often found in prisons and jail locations. They may feel they may not be going to a good place if they try to cross or they still find some twisted enjoyment in tormenting the living and weaker spirits, sometimes trapping the weaker souls, preventing them from leaving or even communicating. They may present themselves as something they are not to gain your trust. They may come off as being a lost child or a sweet grandparent.

Have you ever heard of someone say they "invited" a lost or friendly spirit to stay because they thought they might be just lonely and soon afterwards everything is chaos in their home? Not a good idea. Always keep your guard up, and if there's any doubt, just offer a prayer and guidance to help them move on.

From earthbound ghosts we arrive at ourselves, the living person in this physical world. We are in our corruptible, mortal state, as it were. Our choices and behavior here may determine where we travel to when our physical body dies. If all goes well, we pass into what is formally called spirit.

A true spirit form is the soul free of its physical body. A spirit may choose to remain here to guide and guard loved ones left behind. They have free will to cross over and continue their journey, to become reincarnated into another life. If they had a good life, full of love and wonderful experiences, I can't blame them if they wanted to remain here and relive old times. I believe that once a spirit crosses over they still have the ability to cross back, usually to watch over friends and family. Occasionally, a spirit may pop in and out of familiar locations, places where they felt comfortable and at home. If a person enjoyed this life, they may want to remain in a location to keep an eye on things. This is common in locations that had a dedicated owner; they're so protective and proud of their business or location that they feel the need to continue to watch over it and make sure the living don't screw things up. These spirits remain conscious of who they were on earth and can communicate with us in the present tense. With ghosts and spirits, you are able to communicate, and capture EVPs, or electromagnetic voice phenomena. You can capture them on video or through photographs as well, but it's through mostly photographic evidence we capture the remnants of residual hauntings. Remember they are just fragments, replaying like an old movie, repeating over and over. Residual hauntings don't communicate, intelligent hauntings do.

Beyond spirits we go up the rung to Ascended Masters. Ascended Masters are guides and guardians. Ascended Masters have been in human form and have risen to a pure spiritual form. They are beyond reincarnation. Think saints, Buddha, Gandhi, Mother Theresa. Ascended Masters can also be guides from ancient legends. White Buffalo Woman is one of the strongest spirit guides. She is a spirit elder of the Lakota. The Ascended Masters work as guides and guardians. They assist psychics and mediums to connect with other spirits. I list animal spirits above Ascended Masters because of their unconditional love and devotion. animal spirits often work as spirit guides as well.

Next we go to angels. You don't have to be of a Christian faith to work with angels. You can refer to them as higher vibrational beings if it's more in line with your belief set. Angels guide, guard, and also assist with connection of spirits. Their main job is protection. Everybody's personal angels look different. You may not be able to see your angels at first, but continue to call them in to guide and protect you. After awhile, you should be able to see them. My friend Kimberly Rinaldi's angels look like tall, Lalique Art Deco statues. Mine, on the other hand, look more like Disney's Tarzan. Hey, I didn't pick them, they just showed up looking like that.

At the top is true creation. The Divine, God, collective knowledge, where all knowledge and souls connect.

Psychics learn to recognize and work throughout these different levels of vibration to receive information and connect with spirits. Of course, we can't discuss consciousness and the ideas of life after death without touching on the subjects of religion, science, perception and skepticism. Some folks look for answers regarding the afterlife to validate their faith. Religious doctrines vary on reincarnation, but every religion dictates that your actions in this life will have consequences after the body dies. Whether you call it Karma, Kismet, Purgatory, most folks are inclined to believe we sow what we've reaped.

Some people wonder if all spirits are dead who are trapped and cannot cross into the light. There are those who think a ghost or spirit must somehow redeem themselves in order to be at rest, and that all hauntings are the result of a bad life.

The subject of spirits and ghosts directly leads into discussions of consciousness what makes up our souls. What *are* we? We are basically blobs of energy. So what happens to this blob of energy when the material body dies? The sciences of quantum physics and metaphysics are intertwining and supporting evidence we find in the paranormal field. In fact, the basic law of physics is that energy cannot be destroyed - it lasts forever. We are all energy. So where do we go and can communication between the living and the dead exist?

I wrote this book for several reasons. First, I wanted to talk about what a haunted location really is. What I believe a spirit or ghost is. What makes a place haunted. The thought of dead people still remaining in a location and having the ability to communicate with us to me is not the least bit scary; I find it fascinating and a little bit sentimental. The spirits who can cross back and forth but remain in a location do so by choice. It's the truest definition of devotion for a person who loved a place so much and and had so many memories tied to a place they just want to become a part of it forever. They want their soul tied to a comforting place. Perhaps it was where they met the love of their life.

It may represent their happiest times. It might seem like remaining in a wonderful dream. I wonder if they spend eternity just reliving love, family, the good times. Perhaps it's a sense of duty that ties them to a place. I have encountered workers still toiling at some phantom task, somehow unwilling to stop their work. I wouldn't be surprised if a long dead sea captain still keeps a steady hand at the wheel of his ship. Many ghost sightings are of apparitions in uniform, roaming the area. Are they keeping a watchful eye over us? I like to think so.

I haven't forgotten the souls who may feel trapped and need our help. Indeed, that is one of our group's primary goals. Our group name, Into the Light Paranormal, says it all. We founded the group because we were finding spirits who needed some sort of guidance. Some just wanted to know they weren't forgotten. By making a connection with them and maybe having a conversation with them, they seemed to feel a bit of peace. Just that little bit of acknowledgment is sometimes all they want. Imagine how lonely and confused they must be.

This holds true especially in hospitals and battlefields - areas where sudden and traumatic deaths occurred. These souls may not know they are dead. Everything may look different to them. No one can hear them. These are the ones who need help most. It is always an honor to be able to bring angels in to help these souls. It's an honor and a privilege to be able to assist these souls and give them rest. We have found that once spirit crosses over, they can cross back again and are no longer trapped. By doing so it does not create a void in a haunted location. It has not been my experience that crossing over spirit makes a place any lass haunted. Indeed, I think it creates a type of vortex. Spirits somehow know there are people on this side who want to communicate and are offering help and healing. Word kind of gets out and spirits not attached to a location start showing up. It creates a type of network with the deceased. That's nothing short of amazing.

I also wanted to give the reader a short history of the Spiritualist Movement and a history of communicating with spirits. Knowing how we started connecting with spirits is important. The paranormal field is gaining in popularity. Once completely dismissing the paranormal field as a pseudoscience, traditional science and technology now have given credence to some basic paranormal theories. Acceptance as a credible scientific field is what the paranormal field strives for with each investigation.

Second, I wanted to tell you about some of my paranormal investigations. We travel to some of the most haunted historic locations in America. We use the latest equipment available and we connect with spirits using our psychic abilities. When I first started doing paranormal investigations, I assumed the only way to be a skeptical - and therefore credible - investigator was to base all my findings by what I collected using equipment alone. Evidence was either photographic or audio or a fluctuation in measurable energy in a location. I thought that was all there was.

But as I spent more time on investigations, I began to notice differences in how some places felt. I would get suddenly queasy, or feel anxious. Sometimes I was overcome with a sudden sadness. These feelings would vanish as soon as they came, like bubbles of emotion. I realized I was feeling <u>their</u> sadness, <u>their</u> emotion. I might go into a room where someone had been shot and feel a sudden, sharp pain in my body. As hard as it was sometimes, it was the first step towards true connecting and therefore healing.

This book expands on the idea of using your own psychic abilities during a paranormal investigation. To connect and to feel what spirits are trying to say. I would rather connect by feeling their pain, than be indifferent to their suffering.

Lastly, I will take you through some of my investigations. I'll share what we encountered and what we felt. I'll share what audio evidence was captured and when we caught photographic anomalies and what personal experiences we had. But then we'll review each case, step by step, and I'll guide you on how to incorporate your psychic senses to connect with the energies in a location. By psychically connecting to the energy of a place, it can help steer the investigation to the most active areas as well as communicate with whoever might be there. I think incorporating your senses - whether or not you consider yourself a psychic - will enhance any paranormal investigation. Think of it as adding another set of tools to your arsenal. Your senses and how you feel when you enter a haunted location are very valid indicators of paranormal activity. I want you to feel and connect with a haunted location, to resonate with its history and hear its stories.

I think having all this at your disposal will make you a better and more credible paranormal investigator.

CHAPTER 1:
The Early Years

Why Ghosts?

I grew up in the historic town of Whittier, California, and have never left. Whittier is over 100 years old and has many unique old homes. In fact, the house I live in now was built around 1910. It still has the drain pipe from the old ice box.

My mother was born in Whittier and worked for the Whittier Daily News as a reporter. After World War II, she was assigned to write a story on Marine veterans based at Camp Pendleton. She pulled up to the base, her wavy red hair flowing in the ocean breeze. My dad, a Marine sergeant and veteran of Iwo Jima, was sitting playing cards with his feet propped up on the porch railing.

"You ever ride in a Jeep?" was his opening line.

They were married 49 years.

They bought a little tract house brand new for $12,000 in 1954. This house, for some reason, was extremely active with paranormal activity. Objects would move on their own, and about once a week we would come home to running water out of every faucet in the house.

I can remember as a young girl being home alone and, even during the day, hearing the sounds of heavy footsteps clomping down the hallway. There was something creepy about that hallway, which connected to my bedroom. It was only about 10 feet long, but I just hated walking down it, because it always felt as if a big something was standing right next to your ear.

So I would run through the hallway, just to be out of it.

Of course, when you're a kid, your parents just assume it's your imagination. I even began to consider that myself, thinking maybe it was normal to be a little scared when you're young and alone in a house. But then I noticed that our cats would often suddenly stop and stare in the direction of the hallway - and then run off in the other direction! Not very comforting when you're alone and just 7 years old.

Whatever spirits that resided in the house seemed to have a special affinity for green colored objects. Green things seemed to mysteriously be misplaced, and would subsequently be found in another, odd place - like on top of the door sill, or under the sofa. We found a green drinking glass in the shower.

One summer day, we had plans to go to Disneyland. The night before, I had placed my little green plastic treasure chest on my side table and went to bed. The next morning, we searched everywhere for that treasure chest. I was crying. My parents were upset because we were running late. The morning was going downhill, and fast. Until we found it - on top of the refrigerator. I was much too small to have put it there myself, and my parents surely didn't do it, especially not if they were then going to spend all morning looking for it. So who put it there, and why?

Silly ghosts.

I was never able to find out why our house was so active. My parents were the original owners. Before the house was built, the land was acres of orange groves.

Although I was sensitive to all this paranormal activity, it didn't upset me. I was a little kid at the time. I presumed all houses had some kind of haunting or ghostly activity. It didn't seem weird to me at all. (Except for the hallway. I still hate hallways.) My parents never made a big deal out of the poltergeist activity, so neither did I or my brother or sister.

My sister was truly frightened only once. She was about 8 at the time, and she just had that *feeling,* you know, when something is watching you in the dead of night. She peeked over her bed covers and peered into the dining room. There, floating along about 5 feet above the floor, was a green, disembodied head. She hid her own head under the covers for the rest of the night.

I think I only had one encounter in that house that was too close for comfort. I had gone to bed and was waiting for my two cats to join me. I felt a thump on the bed, but it wasn't that familiar pounce a cat makes when they hop on the bed. I could feel the pressure on the corner of the bed. I could feel the weight of the corner sink. *That's not a cat* ...I sat up in a start. I could see the corner of the bed slowly sinking down as I still felt the weight of something sitting there - something I couldn't see. A spirit was sitting on the edge of my bed, making the corner of the bed sink down with its weight. I didn't know what else to do, so I yelled at the invisible visitor.

"Stop that! Don't do that!"

Instantly, the weight was gone. And they never did sit on my bed again.

Growing up, if you saw a ghost or knew a place was haunted, you just acknowledged it and carried on.

I didn't really try to have conversations or any communication with Spirit until I was an adult.

I was thrilled when I learned that ghost voices could be recorded. I started recording spirits on digital recorders about 25 years ago. To discover that someone who has died can still communicate is nothing short of mind-blowing. The first time I played back a recording and heard the breathy whisper of someone who was dead, and had no body, was simply amazing. To record an electronic voice phenomena, or EVP, will test your religious beliefs, your idea of mortality, and your idea of what time is. You never forget your first EVP.

I attended a business conference aboard the Queen Mary many years ago. I had heard the ghost stories and wondered if I could capture any evidence of these ghosts of the ship. Armed with my trusty digital recorder on the RMS Queen Mary, I set out to investigate the Isolation Ward.

I bumped into a couple of other people who had the same idea. We hit record at the same time and together we asked a few simple questions. I asked "What year is it?" When played back, the response was a whispered "1943." At this moment, I was hooked.

Since then, through diligent research of locations and sometimes simply experimenting with different pieces of equipment, I have developed an affinity for collecting audio evidence more than photographic evidence. A shadow can often be just a shadow, and an orb is usually a dust particle, but a clear "hello" recorded in the white noise is harder to dispute. I have found that EVPs, or electronic voice phenomena, sound different than someone talking in the room. Yes, you still need to make sure no one is in a nearby room or outside as this can contaminate a recording. But a good EVP will sound very breathy, and usually they sound very close to the microphone. It can be a little scary to think they are right there in front of you and you don't even know it.

On investigations, I primarily will focus on how an area feels and let my senses guide me. Sometimes a particular room that has reports of being extremely active may be quiet on a given day, while walking down a simple hallway may send your hairs up on your neck and you can just feel the spirit energy. That's why it's a good idea to walk around a location, and get a feel for it. If it feels like static electricity and your emotions race unnaturally, then get to work. By the same token, even if you feel normal, try speaking aloud and introduce yourself. Try a short-burst EVP session, recording for one or two minutes, and then playback to see if you hear anything. Take several photos and see if anything shows up. If you don't capture anything, come back later and try again. You'd be surprised how the spirits just seem to be watching you and sizing up your intentions.

I love historic locations. They make you feel like you have just stepped back in time. I love the smell of an old building. The wood, the ironwork. You can see the craftsmanship and pride that goes into building these places. I think appreciating and feeling their pride goes a long way in connecting with spirits.

It's important to research the history of a location, read about who lived there. It's about having that background, that connection. I think it creates a sort of mutual respect for a location that the spirits can recognize and hopefully appreciate. Then when you do an investigation and you get an intelligent response from a spirit, it's as though you are adding pages of history to this location. You are now writing pages and adding bits of history with someone who may have died a century ago.

I also adhere to a strict code of ethics during all investigations. I have to credit this to Adam Blai, a clinical psychologist. Blai works alongside the Catholic church as a certified demonologist. While attending one of his lectures, Adam Blai instilled a moral reasoning behind paranormal investigations. He made me realize that I see many so-called ghost hunters go into a location with little or no sensitivity to the souls who were lost there. They don't take into account the pain and trauma people felt in these places. Many times I have seen people go into a sanatorium, orphanage, or hospital only to prod and provoke spirit for "evidence" they can post on their website. Conversely, my years of investigating and gathering evidence has given me a connection with the dead. We are looking to communicate, after all.

At some point every paranormal investigator will probably get some evidence that is a plea for help. What should you do? What can you do?

My goal as a paranormal investigator is to gather evidence and hopefully make contact with the other side. I seek answers to questions I am afraid to ask. I want to know what happens when our body dies. Who are these people? Can they see us? What do they want? Can they want for anything? How can I give them comfort if they are confused, afraid, or in pain?

I always say a prayer for the dead, in case there are those needing help. This prayer can be any sincere, heartfelt plea for their loved ones and the Angels to reach out and guide them home. I have heard recordings of such prayers and when played back you can hear the spirits actually saying thank you.

CHAPTER 2:
Moving into Mediumship

After years of paranormal investigations and fumbling with communicating with spirit, I intuitively developed skills of mediumship. I noticed a big change after one of my visits to the Queen Mary.

I was at work the next morning, and it was very early - only a couple other people were there with me. I was going about my business when my co-worker, Julie, who I later discovered was very sensitive to spirit, says, "Kitty, who is that lady with you?"

I look around. I'm by myself. "That lady, she is all in white. Don't you see her?" I look around. Nope.

It did feel a little funny on my left side, like someone was just there a second ago and I could still feel the bubble of energy they left. Humoring Julie, I asked, "What does she want?" Julie cocked her head to the side for a moment. "She says she wants to thank you. She says you prayed for her."

It took me a second to remember that I had said my Prayer for the Dead on the Queen Mary. Could it be that simple? That obvious? I told the energy blob she was welcome and went about my day.

The next day, five of them showed up.

I began to make out shapes. I began to see colors. Not knowing how else to proceed, I started a game of 20 questions. I began to hear the answers in my head. Maybe it's my imagination, I thought, or wishful thinking. I just went with it and kept asking questions. Through this crude line of questioning, I start to communicate.

By using this method I was able to speak to who I believe was a sailor on the Curacoa, the escort ship that was tragically struck by the

Queen Mary during wartime. It began with me seeing the color blue in front of my eyes.

“Does blue mean water?” I hear a yes.

I see red. “Does red mean blood?” I hear a yes.

“Were you on the Queen Mary?” I hear nothing.

Then I hear “Cur … Cur ..” “Were you on the Curacoa?” Yes.

“What is your name?” I hear Michael. But then I also hear John.

“Is your name Michael or John?” It’s not clear.

Then it’s gets weird. I can start to see legs. His legs. My legs start hurting. He’s in pain. What do I do? I run over to Julie. She sees him too. “He hurt his legs,” she says.

“What do I tell him?”

“Tell him he doesn’t hurt anymore.”

That won’t work, will it? I close my eyes and focus. “You aren’t in pain anymore, you are fine. It’s over. No more pain.”

Poof, he was gone.

That night, I looked up the casualty list from the Curacoa. There were no casualties by the name of Michael, but there was one by the name of John Meikle. Was he the one?

That was the beginning of true mediumship and the validation that I can communicate mentally with spirits. It basically means I can hear spirits speak to me and can converse either out loud or answer them with my thoughts. Mediums are also sensitive to spirits’ emotional states and may be able to feel their pain or fear.

I think everyone has psychic abilities. It is just a matter of developing these skills through practice and meditation. The dead know much that we do not and have so much to teach us; they can answer so many questions if we wish to listen. Everyone has loved ones and spirit guides who watch over us, protect us and continue to guide us in our daily lives.

Being able to bring messages from loved ones who have died is truly a gift and a blessing. I am thrilled when Spirit trusts me enough to channel through me and allow me to bring messages to their family.

My dear friend John lost his father. I meditated that night and focused on reaching John's dad. I didn't even know his father's name, so I simply said, "Please let me see John's dad." I guess they are pretty smart because after a few minutes I started getting images in my head. I could see what I believed to be his dad flying through the sky like Peter Pan. It was dark but full of stars. He was met by a strong man with full arm tattoos. he seemed very much in charge. He took him by the arm and off they flew. Two women, who looked like they could be sisters, joined them. Flying in the darkness, he kept turning back and crying.

"Why is it dark? Why is he not in the light?" I asked. "He is traveling," I heard.

They stopped and let him rest on what looked like a throne of gold. "He needs to adjust. Let him rest," I heard. I took all this information in, not questioning the answers. Randomly, a visual of a baby appeared. Then it was gone.

I wouldn't see John until the funeral. They had a large board of old photos. I saw one with a big, strapping man with full arm tattoos. "That was the granddad," I heard. I looked at the photo of John's dad in his youth, with big muscles and shirtless. "I look pretty good in that one," I heard his voice say. I smiled.

I watched John and his family throughout the service. As John gave his eulogy, I could see a hand resting on his back, comforting him. The two women in my vision were departed aunts. And the random baby? He had a brother who died as an infant.

Ghost versus Spirit

There are schools of thought about the difference between "ghost" and "spirit." A ghost is said to be an earth-bound being who was once in human form but has not crossed into the light. A spirit can be a person who has crossed, or a being of higher vibration, such as a spirit guide.

Spirits who have crossed have the ability to come back and interact with us. They are not bound by a veil. I will, throughout this book, sometimes interchange the term spirit and ghost. I don't always know what kind of entity I'm dealing with at first and there may even be both ghosts and spirits in a location. With older, historic locations, you may have layers of the deceased. They may have different levels of consciousness, moving throughout a location like silent film clips. I will define what type of entity I'm dealing with in each investigation.

If I find a spirit in distress, fear, or in pain, I do my best to provide some sort of healing and closure. I do this with the use of prayer, and through energy healing. While providing healing on the other side, I realized there was a need for healing of the living.

CHAPTER 3:
Healing with Reiki

Especially in residential hauntings, the living family dynamic is a vital component to the how the energy feels and stabilizes within the house. Stress, boisterous and rebellious teenagers, divorce, even folks on medications can swing the balance of energies and trigger or exacerbate paranormal activity.

Poltergeist activity has been known to be triggered by hormonal changes in family members. If you research histories of residential poltergeist activity, often there are teenage girls living in the house.

So I found there was a need to add energy healing for families to my toolkit. I started providing Reiki energy healing to clients. Reiki is a type of energy healing. It helps balance the body by realigning chakras and restores a body's natural flow of energy. By doing so, a body can do wonders at healing itself.

When I first heard about Reiki, I was a bit skeptical. How can just holding my hands over somebody actually heal anything? Am I really supposed to be able to feel differences in a body's "energy" and somehow know instinctively what might be ailing them? That sounded like quack science and snake oil to me. But I heard it would help me focus on recognizing another's energy field and energy from outside sources, such as spirits, so I thought I'd at least give it a try. I was taking classes on Metaphysical studies and saw many of the same students were also taking the Reiki classes.

I heard their stories on how it helped open them up psychically and spiritually, allowing them to feel and manipulate bits of energy around them. A typical Reiki class will lecture on the origins of Reiki in Japan and the different schools of thought on its ability to work and heal. It really didn't make much sense until we applied to a practical exercise.

The instructor handed us each a spoon. A metal spoon, nothing fancy about them. "Now grasp the spoon with both hands, like this." She placed the bowl end in her left hand and the stem end in her right. She then led us through a guided meditation. With each breath, we focused on the neck of the spoon.

"Match the spoon's vibration to your own vibration, even if you have to simply imagine the spoon vibrating."

Then with a line seemingly right out of the Matrix, "The spoon is already bent, you just have to make it happen."

What? But with each breath, the spoon indeed felt softer, warmer.

"Now I'm going to have you count to 10. With each breath, with each number, the spoon is becoming more and more pliable, ever softer."

One through seven felt little change. But when we reached number nine, I was really feeling the spoon becoming like a soft, old carrot.

"10! Bend! Bend!" Sure enough, that spoon bent and curled and twisted in my hands like a rubber band. Then as suddenly as it "melted," it hardened again.

This was something I could get my teeth into! I could manipulate matter by utilizing a universal energy.

Studies on the metal composition of a psychically bent spoon shows the bent section has changed on a molecular level after being bent. The molecular structure now resembles that of hardened carbon steel - something subjected to extreme heat.

Some of the initiations of becoming a Reiki practitioner are considered sacred, so I will not discuss the training or ceremony involved, but if you have an interest in learning Reiki as a healing modality or if you want to expand your energetic vision, I highly recommend learning from an accredited Reiki Master.

Whether you want to use Reiki for healing, enlightenment, or are simply curious, any level of Reiki training will improve your ability to feel spirit.

You will become more sensitive to an energy around you and you will be able to gauge whether or not it is a "good" energy or if it's more of a "negative" energy. This isn't going to make you some sort of a demonologist, but I do believe it will give you a useful toolset for feeling out a haunted location.

Using Reiki on an individual can open up a spiritual channel. I have had sessions with a client and, while giving Reiki energy to them, started hearing spirit voices or feeling emotions that I know were not my own.

I once held a free Reiki clinic at my veterinarian's office, offering Reiki treatments to pets and people. A nice lady came in with two dogs: a big, panting, happy pitbull, and a small, older spaniel mix. She said the dogs were depressed over the death of her boyfriend, and that they were very close to him. As I was sending Reiki energy to her dogs, all I could feel was that it was she who needed the most healing.

I stopped, looked up, and said, "Your dogs are fine. They say they are coping. If you don't mind, I'd like to give you some healing, if that's all right." She agreed. Standing over her, I could feel her despair, her emptiness. I used Reiki to balance and fill that emptiness with wholeness and light.

Then I started hearing this voice in my head. "Sweetie … Sweetie … Sweetie … SWEETIE!!"

"Your boyfriend ... did he talk kind of loudly?" I asked her. She said yes. Even though I was feeling a little awkward at this point, I knew I had to give her the message he was now yelling in my ear.

"I'm just going to come out and say this," I told her. "He says, 'Sweetie, take care of the kids.'" The woman began to cry. The message brought her great comfort knowing he watches over her and the kids (dogs).

Find different and unconventional ways of exploring what lies beyond this physical existence. I am a certified Usui Reiki Master. Reiki transcends time and space. It heals universally. I have been able to channel emotionally-troubled spirits and utilize Reiki to help heal and calm their emotional state. Once at peace mentally, the deceased can move on in their journey.

The Test of Reiki

I didn't always believe Reiki would or could do anything. I had heard how it could heal and make tumors and illness vanish.

But could I use it on myself?

My doctor had discovered some abnormal cells during an exam and I was scheduled for a hysterectomy. When I got home from the hospital, I decided I was going to heal myself. I knew the organ removal was a necessity, but would the abnormal cells spread and become cancerous? I meditated and envisioned pure healing and loving light enter my body. I told myself out loud, "I am happy, healthy and whole." I refrained from saying the words "cancer" or "don't" or "wish I didn't have." Using positive and present tense verbalization is important when sending words into the Universe. Words are spells, that's why they call it *spell*ing.

On my one-week post-op check up, I asked the doctor if the biopsies revealed abnormal cells anywhere else.

"We didn't find any," she said.

"What do you mean you didn't find any? Is that common?" I asked.

"We *always* find residual abnormal cells," she said. "This never happens."

I was thrilled, to say the least, but the real test to prove to myself whether or not Reiki worked was still to come.

The one I hold most near and dear to me is my Pixie Bob tabby cat, Scout. At 9 pounds, this stumpy-tailed cat is all piss and vinegar. Scout and I have been inseparable since the day I brought him home. Now 16, Scout had developed an enlarged lymph node. The biopsy came back a rare Lymphoma. I was devastated. I cried so hard all that day. Chemo would be too much to handle for a cat his age. What could I do? My faith as a healer had been shaken. Would Reiki help my Scout?

That night, I put Scout on my lap and held my hands on either side of his head. I prayed for guidance. I prayed like I've never prayed. I needed to know if this type of healing was the answer for Scout. I needed a sign that my guides, my angels, could hear me and to let me know I was doing what was best for my baby boy.

As I held my hands next to Scout, I began to feel pressure on the outside of my hands. I felt the weight of another's hands resting atop my own. They knew I needed to feel their presence and their guidance. My angels and guides indeed answered. Scout is doing well and still chasing my 65-pound Bull Terrier.

When Reiki is Funny

Trying to use alternative healing methods can create some amusing moments as well. Not everyone who receives Reiki is completely onboard with the idea of some "Universal Energy" coming down and making them feel better. Sometimes a family member is more familiar with alternative healing and calls me into treat another family member who is, shall we say, skeptical.

I received a call from a woman who was interested in Reiki treatments for her father. Her father, Mel, was 97 years young and looking for new ways to increase his energy. Mel still drove regularly and went to the gym three days a week. They lived nearby so I told them I'd stop by the following afternoon.

A Reiki session is an energizing yet relaxing experience. While receiving Reiki healing energy, you feel a sense of bliss and utter contentment. I like to do Reiki in a quiet setting with few distractions.

I arrived at the house to find Mel was not even home yet. He didn't know he'd been scheduled for a session and was running errands. His wife let me inside, assuring me he'd be home soon.

He didn't arrive next, but the gardeners did. Blowing and mowing, creating a racket. *Whir, whir, whir, bbvvvvv, bbbvvv* … The noise from the gardeners alerted the family chihuahua, who apparently didn't like gardeners - or anyone in the house, for that matter.

Yap! Yap! Yap! Whir, whir, whir! I tried to greet the dog, but he was having no part of an intruder in his home and promptly bit my ankle. *Yap! Yap! Yap! Whir, whir, whir, bbvvvvv, bbbvvv!* They put the dog away in another room while I waited for Mel to come home.

Finally, Mel came home and we got started on a nice, relaxing … *Whir, whir, whir!* ... Reiki treatment. The sequestered chihuahua is still mad and is barking and growling under the door. And oh, yes, Mel forgot to wear his hearing aids today.

"Who's she?!" Mel asks, in between growling noises from the dog.

Whir, whir, whir!

"Dad, she's here to give you Reiki." More growling.

"Hi," I try to say. "I have a release form for you to sign. It's because I'm going to place my hands on you during this session."

"You're gonna touch me?!" Mel yells. "Am I gonna be raped?"

"No," I assure him. "That's a different set of paperwork."

Yap! Yap! Yap! Whir, whir, whir!

I managed to get him seated and I don't know why at this point, but I popped in my soothing CD of Tibetan singing bowls and hit Play. He can't hear it and … *Whir, whir, whir, bbvvvvv, bbbvvv!* … neither could I. I showed him the different crystals I use and he was able to feel the difference in the vibrations. He enjoyed holding the stones, switching them from hand to hand, and it kept him busy while I worked on his aches and pains. His neck was stiff but other than that, for a man of 97, he was pretty balanced. I worked on him for a good half hour (I'm pretty sure he dozed off) and was able to break up the pinching in his neck and shoulders. I began packing up my crystals and gear while he dozed a bit.

By then the gardeners had finished and they let the chihuahua out into the room. He sniffed me suspiciously. Mel blinked and sat up.

"I suppose now you want money?!" Mel asked.

Really I just wanted to go home at this point, but yeah, I wanted to be paid too. As we were settling up and I was chatting with his wife, I tried to end my visit with a sales pitch.

"Would you know anyone else, perhaps at your gym, who might benefit from a Reiki session?" I asked.

Mel sniffs. "I don't know anyone who would be dumb enough to sit through that!!"

"Well," I said, "if you *do* know anyone dumb enough, here are some of my business cards."

I do my best to create a bridge of awareness, a connection to people I meet, and the spirit world. It's not a scary place to be avoided, but a place of wonder and questions yet unanswered. I have seen and experienced amazing things. I have a hunger for knowledge and the curiosity to look for answers. I do my best to help those I meet along the way and wish to awaken a curiosity in others. I look for the truth.

CHAPTER 4:
Are You a Skeptic or Believer?

Are you old enough to remember the cartoon Scooby Doo? On every episode, a group of kids and their talking Great Dane find themselves on an adventure. They always encounter some mysterious and scary fiend that frightens people away from a location. Every episode ends with the kids discovering that it's not a gruesome ghoul at all, but some unscrupulous businessman in a disguise trying to scare folks for his own financial gain. Even though they and the whole town were spooked, these kids concentrated on the facts before them and persisted in digging up the truth.

Scooby Doo taught us at an early age that being a skeptic is a good thing. But being a skeptic can be a double-edged sword.

The scientific field still does not consider paranormal research a true science. They've labeled us a pseudoscience based on the fact the majority of our work is field research. The Empirical method is their mantra. To them, a science is based on gathering observable, consistent data, basing a hypothesis and testing this hypothesis through controlled, repeatable experiments. To some scientists, the whole idea of the paranormal is hogwash.

But let's look at an example of the scientific research method.

Skeptics

By 1898, Nikola Tesla had toiled several years with the idea of transmitting electricity wirelessly. His studies led him to to recognize three important necessities: first, to develop a strong transmitter capable of tremendous power; second, to perfect the means for isolating this transmitted energy; and third, to discover the laws governing the propagation of currents through the earth and through the atmosphere.

After building a transformer with recording devices, Tesla observed unusual findings in the recordings whenever a big electrical storm passed through the area. He was able to observe and record the ebbing and flowing of electrical readings on his devices. These observations led Tesla to hypothesize that the planet behaved like a conductor of limited dimensions. He knew proving this hypothesis would lead to creating the ability to send telegraphic signals wirelessly across great distances.

There's a lot more to this experiment, but you get the general idea. Tesla observed a phenomenon, then built a set of instruments to measure this phenomenon under ideal conditions. He created a hypothesis to explain the phenomena and used his instruments and laboratory conditions to repeat it. By controlling variables, Tesla was able to predict an outcome.

We all know that paranormal activity cannot be replicated in a lab … yet. We just don't understand all the variables involved that come together to produce what we call paranormal. That's why it's still considered supernatural. Even the strongest skeptic knows that traditional science isn't static. It's a growing science. Science changes when new information comes to light. It reevaluates its findings based on this new information. Scientists once thought the sun orbited Earth, and we sat at the center of the Universe.

Believers

The other side of the coin is the unquestioning believer. So stoked to be on a ghost hunt, they squeal with excitement every time a door creaks or a wind howls. Every little noise must be something other-worldly. Every flickering light just has to be someone trying to communicate.

The problem with believers is they tend to be even more rigid than skeptics.

The religiously devout will twist any supernatural phenomena to validate their belief set. Some folks will go into a haunted location with a preconceived mindset. Their perception is clouded by preconceptions.

Think about it. We're already calling it a "haunted" location. It's "known" to have a lady in white float down the stairs. How many people do you think will come back from an investigation having captured this "lady in white"? While an investigator should research each location so they don't enter a place blindly, too much research can muddy the experience. Good investigators gather the evidence on the day and debunk in the moment. It's always better to check the area and debunk for audible or visual contamination at the time than to later try to remember every open window or recall which area had ambient street noises.

While I always appreciate the enthusiasm of believers, without research and some measure of detective work, their "evidence" is rarely credible.

Be Somewhere In Between

Never lose the unbridled enthusiasm and wonder that comes with seeking answers. Don't smirk with your arms crossed, rolling your eyes, when a psychic describes a connection with spirit, simple because you haven't captured something amazing on film. It's alright to find something you don't understand as questionable or mysterious.

Science does not dwell on absolutes. Science and true scientists adhere to a known "book" of knowledge and slowly add pages to this book with each new discovery. If a new discovery is made that negates an old theory, that theory is replaced and new experiments are based upon this newly discovered data.

But I believe you cannot go through Life completely left-brained. If this were true, we'd all be great mathematicians. But our brains need to distance themselves away from cold logic and just *feel* the situation also. You need to use your gut instinct. Our survival dictates our brain sometimes needs to make decisions based on what feels to be right. Don't lose that.

When a scientist wants to conduct an experiment, he or she is doing so because they have formed a hypothesis of a proposed outcome. They believe their experiment will produce a specific outcome. So their theory is kind of a leap of faith, so to speak. They believe it will work, and then they go about to prove it. This is the balance I want you to achieve when studying the metaphysical. Enter with no preconceived notions. Never lose your sense of sheer wonder. Believe that connecting to something bigger is truly a spiritual experience.

Just be open minded and keep asking questions.

We'll go through some investigations and I'll guide you on expanding your ideas of what you can experience and how to decipher if what you're experiencing is indeed paranormal. I'll also show you how to use the psychic abilities within you, or if you prefer, how to best utilize a more experienced psychic during an investigation. I'll give you tips on how best to incorporate these skills to direct an investigation toward the most active areas. By picking up on the energies of a location and finding where the spirits are hanging out, you can collect better evidence. By getting in tune with a location, you'll learn when to take photos and when you should start an EVP session. All of these pieces will, in turn, make for a more successful investigation.

The Fox Sisters

CHAPTER 5:
A Short History of Spiritualism

Beginnings

Humans, as a species, have carried out formal burial rituals dating back at least 50,000 years. The idea of invoking these rituals signified the importance of caring for the deceased and indicated a belief in some sort of afterlife. Early man took great care to ensure the body was preserved, either through embalming or mummification. The essentials in life were also packed for the journey. This included food, beloved pets, and sometimes faithful servants were sacrificed and sent along to assist in the journey. Even in cultures where cremation was the preferred method of dealing with a body, a ceremony and gifts or flowers always embellished the occasion.

The question is why? What gave man the notion that there is something beyond this existence?

From early history our ancestors must have believed in something beyond this physical world. The care to ensure a proper sendoff is indicative of their belief that they were *going somewhere*. Oral history, Folklore, and even the Bible describe accounts of spirit guides, angels, and ethereal beings visiting and bearing messages.

Civilizations have long thought they had means to communicate with the other side of the veil. The Greeks had their Oracle of Delphi. The Oracle was consulted for all things important and dictated when to wage war and when to plant crops. Native Americans have long held traditions of sweating out impurities to raise their vibrational state to connect to the Great Spirit.

The origins of what I consider modern Spiritualism first started with scientists seeking to document existence of an afterlife. How does one go about "connecting" with an afterlife and how do we measure it? We had little idea of how the mind worked and the inner workings of the mind. Franz Mesmer (1734-1815) utilized a technique that became known as Mesmerism that would induce trances and altered mental states. This later became known as Hypnotism.

There seemed to be tantalizing glimpses of some sort of communication with people who had passed. The communication seemed too one-sided, us asking all the questions and little precise information being received, at least from an onlookers point of view. That all changed in March of 1848.

In the year 1848, something talked back.

Sisters Margaretta and Kate Fox lived in the small town of Hydesville, New York. Their parents John and Margaret had purchased the farmhouse only a few months prior when the family began noticing odd noises during the night.

Strange taps emanated from the walls nearly every night. So loud were these bangs and thuds that the Fox family barely slept. Footsteps could be heard shuffling down the halls almost nightly. First afraid, then desensitized, the Fox sisters soon became amused with whatever or whomever was rapping on the walls. According to a written account by Margaret, the evening of April 4, 1848, began with its orchestra of taps and bangs.

Curious at this point, Kate and Margaretta wondered what would happen if they could make this into a game. The younger Kate called out to whomever, or whatever, it was: “Do as I do.” She then clapped her hands several times. The knocking answered with an equal number of raps on the wall. A response! Soon the girls were enthralled with this strange game of copycat. They would clap, the wall would answer back on demand. Hearing all the giggling and rapping, Mrs. Fox came to see what the girls were doing. Mrs. Fox devised a way of communicating using yes or no responses. One rap for yes, silence for no. Actual communication had begun.

But with whom? Using their technique and asking many yes or no questions,, the Fox family determined this was a departed spirit of a man who had died on the property. Neighbors were summoned over to witness this phenomena and word spread fast. The Fox sisters became a sensation, touring and demonstrating to large audiences for years. Health and money issues plagued them over their career and at times the Fox sisters felt pressured to confess the contact was a hoax, claiming they learned to snap their toes to make the tapping responses. These confessions were always later recanted, however, and the Fox sisters are still credited with some of the first documented spirit communication.

In the years following the popularity of the Fox sisters came countless others. Demonstrations by showmen claiming to make contact through seances and trances became popular entertainment. As these performances became more profitable, they also became more theatrical. Showmanship was an important aspect of Spiritualism, driving it into the mainstream and audible and visual evidence commanded top dollar.

With money and theater comes fraud. Fakes were prevalent and quickly preyed on the vulnerability of the grieving. Still, the Spiritualist Movement, driven by curiosity and science, continued to make gains. With the incomprehensible tragedies of the Civil War, there was more of a demand than ever to try to speak to the dead. So many families had loved ones leave for war and not return. The country was grieving more than ever.

It was well documented that Mary Todd Lincoln, grieving the loss not only of her husband President Abraham Lincoln, but also her son, frequently held seances in the White House.

Blind faith and fat wallets laid a path for charlatans and cheats. The bereaved and lonely were easy prey for magicians and showmen too eager to believe.

This made life difficult for those truly embracing Spiritualism and experimenting with ways to contact the dead. Seeking answers through meditation and channeling wasn't nearly as showy as a table flying across a room. Real, practicing mediums were pushed out of business by a wave of performers who could "magically" cause a loved one to come forth, complete with table raps, candles flickering, and even some fake ectoplasm for dramatic effect. People were mesmerized and crowds flocked around the world to see these charlatans seemingly conjure up spirits at will. Some were known to tie strings attached to bells onto their toes so they could make the underworld respond at the wiggle of a toe. They would soak handkerchiefs in milk and then partially swallow it only to have regurgitate it as "ectoplasm." The more bell ringing, table tipping, and trumpet playing from beyond, the bigger the crowds and the better the payday.

Deception was so widespread in the late 19th and early 20th century that it was a catalyst for ending the lifelong friendship of two very famous men. Sir Arthur Conan Doyle, creator of the most left-minded and practical detective Sherlock Holmes, was considered one of the most diehard believers in the Metaphysical realm. He considered every psychic authentic and worth their salt. The early death of Doyle's son propelled him headlong into the world of Spiritualism. He promoted seers, mind readers, table tippers, even published a book said to be dictated by his wife's spirit guide, Pheneas.

During this tumultuous time there was another famous performer making his mark. Harry Houdini was the world's foremost illusionist and magician. Knowing how magic tricks are done is the key to spotting a fake spiritualist. Houdini was the best and he prided himself on being the scourge of hoaxers and frauds. Indeed, Houdini was the president of the Society of American Magicians, and was keen to uphold high standards in his profession and expose fraudulent artists. He would sue anyone trying to copy his trademark escapes.

Doyle and Houdini met in 1920. After seeing Houdini's astounding performances, Doyle was convinced Houdini must possess "the Divine gift of Dematerialization." He couldn't believe his friend was humanly able to perform his amazing acts.

Famous magician Harry Houdini could see through many of the deceptions and slights of hand by fake psychics. Indeed, for a short while, Houdini himself worked as a psychic using his skills as a magician.

Houdini was furious seeing so many frauds taking people's money and preying on their grief. He would often heckle con artists in the middle of their performances, standing up from the crowd and proceeding to expose how their act was simply that. Houdini published his exposés in the Scientific American. This made him a very disliked man indeed. And this greatly upset his friend Doyle. This eventually drove a permanent wedge between the two men. One could never sway the other to his particular point of view.

The early years of Spiritualism and finding modern ways to connect with the dead polarized many relationships. Perhaps it was just growing too quickly from a war torn country and the vulnerable were too quick to believe. It was a bumpy start but it gained momentum. People wanted answers. They needed to know if their loved ones were at peace.

Psychics still connect with spirit using the age old methods of meditation, and raising their vibration to become nearer to the spirit realm. In the past few decades many new discoveries have led to new methods, new tools, and even more evidence. We now know spirits can manipulate energy and magnetic fields. Engineering pioneers are creating new tools to help us measure these fluctuations, giving scientific validation when paranormal activity is present.

Science will prove what psychics already experience.
That's the future of the paranormal field.

CHAPTER 6:
Ouija and Talking Boards

Years ago, people who wanted to communicate with the dead but didn't have the skills of a medium needed some way to get them there. They needed tools to help them communicate.

Talking boards, or witch boards, were first mentioned in print in 1886 in the New York Tribune. The article described a rectangular board with letters and numbers, which could be used to spell words and messages. The article never specified that these messages were emanating from the "spirit realm," but it did imply something "witchy" was at work.

Enter the Kennard Novelty Company. In May of 1890, the Kennard Novelty Company applied for Patent No. 446,054 for a Ouija Egyptian Luck board.

The Kennard Novelty Company was founded by William Fuld, Elijah Bond, Harry Rusk, Colonel Bowie and Charles Kennard. All these men were members of the Order of the Free Masons. The odd symbols that are still on today's Ouija board are well represented within the Masonic order.

The first Ouija board sold in 1891. It was a stenciled board, or plank of wood. The pointer, or planchette, was also made of wood.

Using a Ouija board and having Ouija parties was a new and exciting way to interact with other young, single people. It involves men and women sitting closely together, sometimes knees touching, in a dimly lit room. It became popular for many reasons.

Several versions were manufactured in the early years. Isaac Fuld created a similar Oracle board around 1901. Some boards had exotic names such as Rajah, Swami, and Yogee.

Even an electric model made an appearance, but was deemed too dangerous.

I have a small collection of vintage and antique Ouija boards. I am fascinated by the history and its connection with the Spiritualist Movement. But they are a tool. And as with any tool, if you don't use it properly, you can get hurt.

I do not use Ouija boards. I do not recommend using them. To me, using a Ouija board is akin to making random phone calls and telling anyone who answers to come on over.

I've never had anything productive come out of using a Ouija board. It has no boundaries. If you want to use them, that's up to you.

Just don't say I didn't warn you.

CHAPTER 7:
Living History

We live in such a technology-based world today. We want everything in an instant and we want it at our fingertips. We have social media everywhere. I'm not even sure how some of them work, but these sites allow us to send and receive news as soon as it happens. We're always on the cusp of what's happening right here, right now.

But sometimes the more technology I use, the less connected I feel.

How many of your Facebook friends have you actually met in person? You wonder if some of them are even real. We know all of these people and know all of these things, and yet where do we come from?

Yes, we need technology and science to build tomorrow. But I feel something is missing. There's an emptiness. We need to understand the past to build a future.

There's just something I get when I walk into an old building that you can't get online. Something rich and tangible. To just enter with an open mind and use all your senses to drink it all in. The buildings speak to you. There's a welcomeness and sense of belonging. Sometimes I leave my equipment behind and use my body as the instrument. I soak up all the energy, the stories, and become a contributing part of a location.

Let history speak to you, and it will.

When you bring equipment and get evidence, you do get validation of what you're feeling. But it's important to allow yourself to feel it. We are often so caught up in validation that we neglect to hear the stories these places have to tell. Without knowing some of the history of a location, you're going into an investigation blind and ignorant.

Do your research on the location, but be open to experiencing what the energies have to offer. I like to think that by communicating with the spirits in a location, we're adding new pages of history.

To be on a paranormal investigation and connect with those who lived and died within the walls is like reopening an old, dusty book. To me it's a bit like time travel. I love to walk on floors that many generations have walked on; I connect with the energy.

It gets even better when you get that evidence, that validation, of what you're feeling. From the scientific standpoint, feeling energy is good - gathering proof of the afterlife is better! To hear in their own voices - through EVPs - what they went through, how they lived, is like a living history through their own voices. That's what makes paranormal investigations so special to me: hearing from the people who lived it.

Some locations hold a special place in my heart. The RMS Queen Mary is one of those places. Every time I visit the Queen Mary I feel welcome. Maybe I was born a century too late. I get such a feeling of comfort, as if I just entered the home of a close friend.

The Queen Mary was built in 1936 and is a marvel of Art Deco splendor. It was pulled into military service during WW2 as a troop transport and was the fastest ship in the world in her day. There were deaths of POWs and soldiers. She carries the energies of decades of parties and the anguish of war. I love her. She's so big that I still occasionally get lost on her, but I don't mind. At every new turn I discover a new place to explore.

I've seen full bodied apparitions aboard the Queen Mary. Guests of years gone by still walk the decks and gaze out from the promenade deck out into the ocean. I've smelled their heady perfume. I've seen shadows and heard conversations of sailors who still toil in the boiler room. We do our best to communicate with spirits who seem comfortable and at peace aboard her, and try to bring comfort to spirits who need help. We will also bring in Angels and healing to areas of the ship where the energy feels oppressive. Conversing with spirits and healing trauma from the past is what an investigation should be whether or not you get physical evidence. You see and hear history with your own eyes, how cool is that?

I did an investigation at a private residence that was very near a railroad. The homeowner would see shadows and hear voices from another room. Her little girl would frequently have conversations with invisible friends, waving at and greeting something the mother couldn't see.

We set up all manner of equipment, lasers, infrared, digital recorders. During our initial sweep, I saw a spirit standing in the corner. He seemed very shy. He looked like a railroad worker, with denim overalls and a cap. He wasn't having any part of our equipment. We were sitting in the room conducting an EVP session. I could feel his energy come into the center of the room. Another investigator noticed the laser pattern fluctuate across my arm. He was near.

"Can you touch this light and make it flash?" He wouldn't touch it. In my head I heard, "They're gonna blow up." He thought the equipment was going to explode. This demonstrates why it's handy to have a psychic along on an investigation. This spirit wouldn't have any part of all these blinking boxes. Being a railroad worker, he was probably familiar with dynamite and to him these newfangled tools looked downright dangerous. He preferred to talk to us psychically and we felt the pins and needles of his energy brushing against us.

If a psychic hadn't been in the room, very little information would have come forth from this spirit. Sometimes the electronics are so unfamiliar with spirits they avoid them altogether. How could a Civil War soldier of Victorian lady know what a K2 is? Sometimes our body is the best tool. It's been around longer and is a familiar vessel for spirits.____

When you are excited to be in a location, the spirits know. I think they pick up on that. After all, if they have decided to hang out there in the afterlife, they must have an equal affection for the place. It's making this connection that will make you a better investigator. Spirits will relate to you better and I think you will get more and better evidence. Having a common bond with spirits is what this is all about.

CHAPTER 8:
Why Do You Investigate?

Ghosts and spirits are experiencing a bit of a renaissance in recent years, even though it's been over a century since the Spiritualist Movement. The steamroller of reality TV brings us all the glamour that infrared and heavy editing will allow in a one-hour format. Folks can begin to think a paranormal investigation is a fun adrenaline rush.

But this newfound influx of researchers and popularity can make it difficult to find dedicated people willing to work for their credibility. So let's take a look at the serious and the curious - and where everyone stands.

Some people just want to dip their toes in the paranormal water, so to speak. Beginners or flashlight-toting amateurs are out for a personal experience. Their approach is similar to what author Jeff Belanger refers to in "Legend Tripping."

In Belanger's book, he writes, "Folklore is about story and oral tradition, paranormal investigation is about trying to apply science to something that can't be measured by current means. Legend Tripping is about the experience and the adventure."

For those who are "Legend Tripping, any experiences they encounter are for them alone. Little documentation or research is done and most activity encountered becomes little more than great stories. Of course, there's nothing wrong in going to historic places to just enjoy the ambience. Historic landmarks and museums thrive on this. At this level it becomes difficult to expand knowledge to a larger skeptic audience, much less credibility with their peers - but it's a fun way to spend a vacation.

Are your motives to seek evidence? To offer help and healing to those on both sides of the veil? To find answers to some really big questions about consciousness, the afterlife, and maybe life's origins? This type of investigator may take a more technical approach when studying locations.

This is what we see mainly in our popular reality shows. We research a location's history. Why might it be haunted? Could the geographical location be causing electromagnetic anomalies instead of spirit activity? Could the materials with which the location was constructed be holding ambient energies?

We do our best to illicit a technical and skeptical mindset while conducting experiments, trying to control as many variables as possible. We base our findings on the data provided by impartial equipment. Paranormal investigating is primarily measuring energy and fluctuations in energies. These investigators do their best to measure something that by its spiritual nature cannot be measured by Empirical methods.

In many circles, paranormal investigators are not considered true scientists, and therefore their findings are not always taken seriously or found credible. They fall into a category I like to call Field Investigators. Not scientists, in the true sense of the word, but doing a good job within their means and abilities. If they are conscientious, I won't be a snob and snub their efforts.

Doing a true scientific experiment involves building a hypothesis, setting up an experiment with controlled variables, study findings, repeat experiment, and report results. This is not always possible when the paranormal is so spontaneous; we don't know everything involved to make hauntings occur.

So where does a dedicated, credible, knowledgeable, amateur ghost hunter stand in all this quagmire and confusion? No one should be ashamed of being a Field Investigator. Evidence gathering and sharing that evidence will drive the paranormal field forward. Evidence gathered responsibly gets us closer to answering really big questions.

I think people investigate haunted places because they are curious. It's what we do during an investigation and how we present what we define as paranormal where we need to be careful.

This book is not for those who just like to run around in dark, creepy places hoping for an adrenaline rush. I wrote this book for the folks who are doing their best to be dedicated, professional investigators. I will show you how I do an investigation and walk you through how to incorporate all your senses into a location. I do not claim to be the best in the field. I know there are well-respected people in the paranormal community who have been doing this longer than I have, and everybody's got their own methods. I want to instruct an individual to use all their senses as wells conduct a thorough and proper investigation. The more tools you have at your disposal, the better your chances for success.

In this book, I'll go through some of our past paranormal investigations. I'll explain why we chose the location, what equipment we used, and discuss our evidence. I will also lead you through each spiritual experience. I want you to feel what I felt. If you are in a similar situation and feel like running out of the room, I'll give you an alternative. I want you to interpret the energy. Is it your fear, or are you feeling someone else's emotions? When you feel overwhelmed, I'll give you the tools you need to focus on what's really happening around you and how to use it to your advantage. I want you to get in tune with your own unique psychic abilities. I want you to trust your best tool: yourself!

The great thing about using your innate psychic abilities is you can go to a haunted location or even a museum and have a paranormal experience. You may not be able to bring in cameras or recording equipment, but that shouldn't stop you from connecting if you find yourself feeling something around you.

We are always looking for new places to hold a paranormal investigation. We'll check out locations during the day, such as a museum exhibit, and if we start picking up on energies or have contact with spirits during the middle of the day, it's a good sign we might get something even better during an evening investigation.

My friend and colleague Kimberly Rinaldi was with me when I was checking out such a place. The former Movieland Wax Museum in Buena Park, California, had been a tourist landmark for decades. The company decided to consolidate the attraction and moved the wax figures to its Hollywood location. But the building reopened with an exhibit of artifacts from the Titanic. Not only do you have the residual energies from all the items from the ship, but the Wax Museum itself had its own bit of activity going on.

We thought we would make a girl's day out, have lunch and see "Titanic: The Experience" exhibit. It was a slow day and we were just about the only people in the building. We really didn't have any expectations.

Before we even entered the exhibit, both of us noticed several spirits were sitting up in the rigging above us. They were just watching us. They didn't want to interact, so we kept going. The Titanic artifacts are all in glass cases, with the exception of a couple of large pieces like the Titanic's whistle. I like to touch the object if I can, to better read its energy. This is what is called psychometry. But that's not always allowed in museums.

We approached a case that held small, personal items, including brushes, a cigar case, and a mirror. At the sight of these, both of us stopped in our tracks. We were overcome with sadness. When we moved to the next case, which was filled with kitchen dishes, the sadness subsided. The dishes were utilitarian and held no personal attachments. We moved back to the original case and felt the sadness again.

This is a good exercise to hone your senses. Trust what you feel and if it feels like it's a bubble of energy, step away from it to see if it changes. Then see if you can step back into it. This will come in handy during an investigation.

So can you be a credible paranormal investigator without having an arsenal of equipment and a team of folks all wearing matching black shirts? I think so. Seeking answers and connecting with those who have passed are personal experiences. And it should be. Don't get into the paranormal simply because you think it would be cool to say you do it. That's the wrong reason for any endeavor. If you visit haunted locations because you like the history, that's fine. If you consider yourself a professionally-minded investigator who approaches a location with an open but skeptical mind, great. If you put evidence out there, put the most credible evidence you've got out there and share how you obtained such great evidence.

Success will unite the paranormal field. When so much irrefutable evidence becomes available and is shared in a central database, the sciences of Metaphysics and Quantum physics will unite and embrace the paranormal. Whether believers or skeptics, we must move beyond our comfort zone to understand how science is subject to social, religious, and cultural influences. We can broaden our understanding of how the universe works and how we are all cosmically connected.

I think the author and poet Edgar Allan Poe said it best: "The boundaries which define Life from Death are at best shadowy and vague. Who is to say where one ends, and the other begins?"

CHAPTER 9:
Earth Energy and Hauntings

"What we have called matter is energy, whose vibration has been so lowered as to be perceptible to the senses. There is no matter." – Albert Einstein

We and everything on this planet and in space are a form of energy. We are all moving and vibrating at different frequencies. This theory was used by Einstein to argue his thoughts on time itself: that there was no linear form of time, and that time as we know it was created just so we could keep track of events and bring a sense of control into our daily lives.

Einstein and quantum basics lead us to believe that time is more like a series of snapshots. You can flip back and forth through them and they become a series of now. Confusing, yes, but it does give us a scientific argument about ghosts and spirits. What if there was no such thing as ghosts? What if we are only seeing bits of history from another time – pages being flipped in some giant photo album? There have been cases where spirits have not only acknowledged the investigators, but seemed offended by these trespassers. EVPs were captured saying they were calling security. Could we be the ones who look like ghosts to them?

If we are all bits of energy vibrating at frequencies, then it makes sense that other frequencies and universal forces affect us in turn. Frequencies affect other frequencies.

Nikola Tesla experimented with machines that could topple buildings simply by matching the frequency of the structure. He nearly collapsed his entire office building during a test!

"If you want to find the secrets of the universe, you have to think in terms of energy, frequency, and vibration." – Nikola Tesla

Tesla said it. Einstein said it. Quantum science is proving it.

The science called cymatics illustrates that when sound frequencies are passed through matter, it directly affects the vibration of this matter. It is best illustrated in soft material like sand or water. Think of tiny BBs on a stereo speaker; turn on the music and they skip and dance. But there have been studies that have discovered the particular frequency of 432 Hz is harmonious with the properties of light, time, magnetism, and gravity. Science even ties this magic equation to harmonizing DNA and consciousness. The number 432 is reflected in ratios of ancient structures such as the Great Pyramids, Machu Picchu, and Stonehenge. Our ancestors were on to something that we are just beginning to grasp.

Another interesting fact is that the frequencies in the light spectrum are related in octaves. These octaves correlate with our body's chakras. Many cultures already intertwine these two together. For example, Tibetan singing bowls are tuned to a specific chakra on the body. When you hold a singing bowl and make it ring, you'll feel it deep within your body. It's designed to stimulate a particular chakra. Buddhist chants hit particular keys to enhance meditation. Some people meditate with special tuning forks; the vibration affects your entire body. We all know that music can sometimes change our moods. That's because musical frequencies can change the vibrations of the fluid within us. We are 70 percent water, so it makes sense that sound can affect us for better or worse.

Dr. Masaru Emoto discusses his experiments with this in his book "The Hidden Messages in Water." He exposed drops of water to different stimuli and then froze them. Studying the drops, he found startling correlations between the words or sounds and the resulting ice crystals. The water-memory theory, developed by Jacques Benveniste, illustrates how the conductive properties within water and the vibrational stimulus of water can change it physically – but I'll get back to that later. If our material world and the universe – and what lies on the other side of what we know – are all intertwined energetically, then we need to explore the other natural forces that affect these dimensions. How these forces are affected directly affects how we see things; it's our perception of what's reality.

So stick with me as we continue down the rabbit hole ...

Electromagnetism is one of the four fundamental forces, or interactive forces. The others are conventionally recognized as gravitational force, or gravity; strong nuclear force; and weak nuclear force.

Because we are able to stand on the ground and not float into space, it's gravity that affects us most profoundly. But a close second is electromagnetism. Electromagnetism is the fundamental force that causes an interaction between electrically-charged particles. An electromagnetic field, or EMF, is involved in the creation of atoms and is an integral component of quantum mechanics. Electromagnetism affects us by tricking our perception. It can affect our spacial awareness, creating feelings of paranoia, nausea, and uneasiness. Electromagnetism is all around us; batteries, magnets, and electric current are all types of electromagnetism. Even the entire magnetosphere that surrounds our planet is a form of electromagnetism. It's what binds protons and electrons together as atoms, and as such is an integral building block for life.

I can't claim to have a lot of expertise in quantum physics, but knowing how matter and forces work and interact is imperative in understanding paranormal phenomena. Ghosts or spirits create fluctuations in an EMF when they manifest.

It's a generally accepted theory that spirits "feed" on energy from an outside source and emit EMF, thus causing the fluctuation. It's this fluctuation of an EMF that most of our paranormal equipment detects. This is why it's important to touch on these fundamental interactions. There seems to be a direct correlation between spikes in EMF and paranormal activity, be it intelligent or residual.

This is why using an EMF detector labeled "natural" may work better than other types.

A natural EMF detector is calibrated to eliminate man-made EMFs such as microwaves, cell phones, and wiring within the house. The Mel Meter created by Gary Galka operates in the range of 30 and 300 Hz and is designed specifically for use in paranormal investigations. Galka's models aren't likely to give false readings like other contractor-grade EMF detectors. I highly recommend them.

Spirits can manifest more readily if you provide some form of energy on which they can draw. EMF pumps are a useful tool. Some people play white noise. Noted EVP specialists Mark and Debbie Constantino like to turn on ceiling fans or will run a water faucet in a room.

Watch your batteries, and you should always carry extra. So many times batteries can be fully charged entering a room, and be drained in moments. Spirits can draw energy from you, too. If you are feeling drained or nauseated suddenly or when you enter a particular area, that could be spirit drawing energy from you. Some people think that's cool and feel they have a special connection if that happens, but I am a bit wary of it and don't trust them when they continue to do it. The spirits may have ulterior motives, so give them something else to draw energy from and be safe.

It may seem like I'm getting off track, but when we're dealing with the paranormal and spirits, we really are dealing with energy and changes in energy. When we capture a photographic anomaly, we're capturing energy. When we feel a drop in temperature, it's likely a draining of energy. If we can understand some of the laws of physics and marry them with quantum physics, we get closer to finding answers.

It is believed that our soul, our consciousness, is pure energy. When our lives here end and we cross into spirit, we become a pure form of energy in the universe. We remain in a state of being – somewhere. This theory is supported by one of the primary laws of physics. Known as the law of conservation of energy, the law states that the total energy in a closed system cannot change, but can be conserved. Energy can neither be destroyed nor created, but can change form or be conserved. For instance, chemical energy can be converted into kinetic energy, like blowing up a stick of dynamite.

So that begs the question, what happens to us when we die? If our soul goes up into the great collective of the universe, does our personality remain intact? Do our memories remain, or are we rewritten into a new plane of existence? Do we simply dissipate like a melting snowflake?

A prevailing belief in many religions is that you are who you are now, and then you (hopefully) ascend to a higher place. In the paranormal community, we seem to have evidence that somehow a person's identity remains intact and may remain here whole, as an intelligent haunting, or as only bits and pictures remain, as in a residual haunting.

CHAPTER 10:
Stone Tape Theory

Part of being a paranormal investigator and researcher involves trying to figure out why some locations become haunted and remain haunted over a long period of time.

The stone tape theory is similar to the imprint theory, the difference being that while the imprint theory's energy resonates from the universe, the stone tape theory states certain types of stone and other natural minerals have the ability to capture and retain information and energy within itself. Much like an audio tape, a building or location can retain the memory of a traumatic event. This theory holds validity when you remember iron oxide is an abundant element on earth. If you think about it, iron oxide is used to make audio tapes – magnetized iron oxide. Could this be why old brick buildings always seem haunted? Our planet and everything on it is alive and vibrating, so it makes sense that a structure made with natural materials from the earth would also somehow be alive, in a way. Brick buildings that sit abandoned for years still have a life within them, a personality.

Quartz crystal can retain energy and information. Silicon is used in the manufacture of computer memory chips and is the second most abundant element on earth. Granite, limestone, quartz – all are capable of holding energy and information. Quartz is harder than steel and doesn't split easily. Electric current pass through quartz. Quartz is used in the manufacture of radios, lenses, televisions, and watches, and have long been used in the manufacturing of granite, a major building material. The Stanley Hotel, one of the most paranormally active locations in America, is built right into the mountainside. Its granite and limestone most surely have an affect on the hotel.

The area of Benedict Canyon in Beverly Hills, California, is notably one of the most active locations in America. The entire canyon is a geomagnetic anomaly. It holds magma pockets and its geographical makeup and behavior does not match other hills of its relative young age. This area was believed to be sacred ground to the local Native American tribes. They knew there was something unusual and special about this area.

There are many tragic and infamous deaths that have occurred within the Benedict Canyon area, including a battle between the Tonga Indian tribe and cavalry and even four murders attributed to the Manson family. One house in particular demonstrates several of the naturally occurring anomalies.

Dave Oman's home is built right into the hillside; the steel support beams and metal staircase inside his home act as a conduit and magnify the energies of the canyon. It's like creating an antenna for paranormal activity. Taking readings with EMF meters, Dave's house can fluctuate from 2000 milligauss positive to over 1000 milligauss negative within the same property. It helps to be aware of these types of fluctuations of EMF on a location because wild fluctuations or unusually high readings can cause nausea, confusion, and dizziness.

There are thousands of prisons across the country and all have undoubtedly had their share of pain and abuse. But the ones that seem to hold this torment and suffering beyond their operating years are the ones made of heavy stone. The ones built of limestone and granite such as Eastern State, Alcatraz, and Ohio State Penitentiary seem the most haunted. Eastern State, built of limestone, had a most cruel method of treating its inmates: "rehabilitation" was continuous solitary confinement. The inmates were allowed little or no human contact, had to wear blindfolds when moved from one place to another, and weren't even allowed to speak. The warden believed this solitude would turn them toward God. Instead, it drove them mad or, in some cases, caused permanent blindness. Years of this abuse was certainly imprinted within the very structure of Eastern State. Visitors today still feel the despair and hear moans and cries throughout its halls.

In 1882 the Sloss Furnace was built in Birmingham, Alabama. A huge producer of pig iron, a major component in steel, this factory helped build America's railroads. Iron permeated every pore of the factory, its grounds, and its workers. Combine that part of the equation with nearly 90 years of some of the most brutal and unsafe working conditions – sweltering heat and humidity, and working alongside huge dangerous machinery caused many work-related deaths when men were regularly crushed by machinery, fell into furnaces, or were ripped apart by gears that never stopped – and you have a recipe for extreme paranormal activity. Apparitions of workers are regularly seen still toiling at some dreadful task. Feelings of dread and encounters with the overbearing boss known as Slag are frequent. Spirits who were controlling in life tend to remain so after death.

That's why knowing some of the background history of past occupants of a location becomes helpful. You'll know how to approach and deal with their energy.

Mine shafts and gold mining boom towns are a hotbed of paranormal activity. Could it be that the stone itself holds the past within it? Gold ore is often encased in quartz. These were also usually lawless towns and violence was a part of everyday life. Notice the correlations of traumatic events and energies with the natural minerals in an area? It makes for a haunted place.

Tombstone, Arizona, and its Bird Cage Theater claimed 26 deaths, mostly by shootings, in its short eight-year span. Tombstone grew out of a silver mining boom and dried up when the mine did. Its local newspaper, the Epitaph, still produces the news in print. The town of Tombstone is infamous for its shootout at the O.K. Corral, but the rather small Bird Cage Theater was the place to be for gambling and other pursuits with their "soiled doves." It also had lively stage entertainment for miners and gamblers. The Bird Cage is also the location of the longest continuously running poker game in history, lasting day and night for eight years. It's estimated that over $10 million was gambled in its poker room. Sounds of cards shuffling and

Morgan dollars clinking are heard regularly within its brick basement. Upstairs in its "birdcages," or lofts overlooking the stage, was where the prostitutes could ply their trade. Here, EVPs of moans and women's voices are often captured. Did the combining of the mineral silver ore and violence of the Wild West produce a perfect place for paranormal activity?

Bisbee, Arizona, is known for its copper and green turquoise deposits. It's filled with quaint antique stores and small local artist shops. Sitting in the middle of this is the Copper Queen Hotel. Built in 1902, it's one of the oldest and most paranormally active places in Bisbee. Rent out Julia's room and the spirit of this prostitute just may tickle your feet.

Ships made of iron almost always contain some paranormal activity. The RMS Queen Mary is very active, even though most of her life was filled with happy events carrying and entertaining passengers. True, she served as a wartime troop transport, carried POWs, and had a deadly collision with her escort ship, the Curacoa. But many of the spirits I meet on the Queen Mary are civilian guests, and they seem to still be enjoying riding aboard the vessel.

On the other hand, iron battleships are floating gems of paranormal activity. Sailors' voices can often be heard along with clangs and bangs of seamen still serving. The USS Hornet is a hotbed of paranormal activity. The Hornet served in both World War II and the Korean War. She served an integral part in the Apollo program, retrieving astronauts when they returned from Lunar landings. Decommissioned in 1970, she now sits as a maritime museum. She is a regular stop for ghost hunters.

A murder, a suicide, a war, a gunshot in time. How do these events become a haunting? Stone tape theory and its cousin imprint theory only partially explain why hauntings occur. Imprint theory tends to describe what we call residual hauntings, when paranormal activity manifests like frames in a movie, indifferent to our presence. Our added amount of external energy seems to give them a stimulus and fuel to manifest, but they don't have enough of a consciousness to interact. They are just memories of lives and events.

Imprint theory has been attributed to recurring apparitions or manifestations. Over and over they play, sometimes in an area as small as a staircase.

One of the most famous ghost photographs is the "Brown Lady," thought to be the ghost of Lady Dorothy Townsend. Facts and rumors blend regarding what was the cause of her demise. Mistress of Lord Wharton, her fiance Charles is suspected of faking her funeral in 1726 and locking her away to rot. The story of how this photo came to be is interesting. It was taken in 1936 by Captain Provand and Indre Shira, staff photographers for Country Life magazine.

According to Shira, "Captain Provand took one photograph while I flashed the light. He was focusing for another exposure. I was standing by his side just behind the camera with the flashlight in my hand, looking directly up the staircase. All at once I detected an ethereal veiled form coming slowly down the stairs. Rather excitedly, I called out shapely, 'Quick, quick, there's something.' I pressed the trigger of the flashlight pistol. After the flash and on closing the shutter, Captain Provand removed the focusing cloth from his head and turning to me said, 'What's all the excitement about?'"

The Myrtle's Plantation in Louisiana has many spirits haunting its grounds, but one of the most notable is the spirit of William Winter. Answering a call at the front door proved to be fatal for Mr. Winter, as he was shot in the chest as he opened the front door. He managed to crawl up the stairs of his mansion where he died in his wife's arms on the 17th step. Moans, thudding steps and misty apparitions are frequently captured on the stairs at Myrtle's Plantation.

Battlefields are a wonderful example of a large scale residual haunting. The cannon fire and shouts of soldiers have left lasting imprints on what was once a farmer's field. The land surrounding Gettysburg was transformed forever. Visit any battlefield and when it's quiet you can still hear the crack of gunfire in the distance and the phantom cries of long dead soldiers.

Stone tape theory gives us a better understanding of the hows and whys of an intelligent haunting, but so many questions remain. How can a residual haunting be in one part of a location – say, a lady in white appears on the stairs – but you can have a lengthy Spirit Box conversation with a spirit in another room? We just don't have these answers yet.

What also remains puzzling are the circumstances in which this energy is released. Weather conditions may be a factor (for example, lightning can cause electromagnetic flux), and the time of year or the lunar cycle all may be factors. Plus, some people may be more sensitive to these energies than others.

Some think energy is released as water retained within the stone evaporates, releasing fragments of an event ... but that's another theory for later.

CHAPTER 11: Water Memory Theory

We have touched on the theory known as stone tape theory, where the argument exists that natural materials such as stone, limestone, silica, quartz, and iron oxide have the capability to not only capture but retain energies from the living world around them. If we are to believe this theory to be plausible, it still only answers part of the equation. If these materials can capture and hold "energetic memories" and provide a possible reason for location hauntings, what could be the trigger for releasing this energy? Is this when we experience paranormal activity at certain locations? What makes the limestone of Trans Allegheny Lunatic Asylum give up its secrets?

There is an extension of the stone tape theory known as water memory. Rocks, stones, and minerals all contain ancient water molecules. What if it's not the stone but this ancient water deep inside these minerals that is actually capturing and recording these events, only to disperse the energy and the connected memory when the water slowly evaporates?

Water is a natural and effective conductor. It makes sense then that energy can pass easily through water, making it a perfect catalyst for the paranormal. It's a scientific fact that EMF is higher near a water source. Spirits need an energy source to manifest, either audibly or visibly. So many locations that are haunted have a body of water in the equation. Why is this? Does the water proximity have a direct relation to the level of haunting in a building?

If you look at a list of all the most notable haunted locations in the world, many of them are near water. Bob Mackey's, a nightclub in Kentucky, has a stream nearby, the Licking River, where the blood of slaughtered cattle was drained. A sign over the entrance reads: "This place is reportedly haunted. Enter at your own risk. Not responsible for demon attacks."

It started with a vicious murder; Alonzo Walling and Scott Jackson stabbed and killed Pearl Bryan in 1896, decapitated her and tossed Pearl's head down the well. Before being hanged for the crime, the two men's last words were that they would haunt the land forever.

Its subsequent life as a juke joint and rum running brought more violence and bloodshed. It became a place known to have satanic rituals performed and blood sacrifices. Demonic attacks from all this dark energy escalated to the point that an exorcism was performed on the entire property in 1994. It's still a negative, negative place. Certainly something haunts Bob Mackey's and it isn't friendly.

The RMS Queen Mary is still afloat in Long Beach. She was the fastest ship in the world in her day and is one of the best examples of Art Deco splendor. She was drawn into military service as a troop transport because of her speed. Because she was so fast and so successful at moving troops quickly, she had a bounty placed on her head for anyone who could sink her. Her escort ship Curacoa accidentally crossed her path and was cut in two by the Queen Mary. Several hundred lives were lost. She is an iron ship still afloat, and experienced wartime. These factors all contribute to her being very haunted.

Gold and other mineral mines are also hotbeds of paranormal activity. Life was dangerous and isolated for men working the claims. Gold miners often would utilize streams to sluice or separate the valuable minerals from the lighter weight rock and silt. Could the stream beds that carried their hopes and dreams still contain some of the memories from Gold Rush days gone by?

Mine shafts and quarries are notorious for being active. Water and natural minerals are the common threads for these industries. The planet is alive and water is a major component of life as we know it. Does something happen to water when it is exposed to traumatic or other emotional events?

There are several theories that water can absorb and change when exposed to expressive energy, traumatic or blissful. In the late 1990s, Dr. Masaru Emoto conducted water messages experiments and discusses them in his book "The Hidden Messages in Water." Dr. Emoto took a single drop of water from different sources, a pond, a well, tap water, glacier water, and subjected these drops to different stimuli. One was blessed by a monk. One was told loving and caring thoughts. Then others were cursed at, told hateful things and verbally abused another. Then these drops were frozen. When viewed under a microscope, the drops that were blessed produced stunning and symmetrical ice crystals. The abused ones became misshapen and irregular and sometimes discolored.

People are 70 percent water. Think about what happened to a single drop of water. Our bodies can manifest physical and physiological change when certain stimuli triggers a memory. War-torn soldiers jump every time they hear an unexpected bang or a car backfires. Mothers instantly respond to a baby's cry. Why, under certain circumstances, couldn't the water molecules within stone somehow react and release their stories? Could it be that as this water slowly disperses, the energy and stories from the living disperse as well? Could it be the reason we rarely experience ancient ghosts?

While places such as Stonehenge and the the Mayan ruins are truly spiritual, have you ever heard of a ghost of a Roman soldier or Neolithic Man? Do our spirits fade?

With every theory comes even more questions. As we delve ever deeper into what is considered paranormal, I hope we can find more answers.

CHAPTER 12:
The Human Body: The Best Investigation Tool

I can feel a connection with the history of a location soon after I step onto the property. The RMS Queen Mary is my favorite and where I most feel at home. Places within the ship have a distinctive feel. I recognize how these places on the ship feel and quickly notice any differences in familiar areas. This is why I think it's so important to investigate locations many times and over the course of years, if possible. As I become a more experienced paranormal investigator I can gauge the energies of a location more accurately. I can tell what's only an old house creaking with decay and age and what's something otherworldly. I am becoming more aware of my surroundings – not only during paranormal investigations but in the physical world as well. Honing your own psychic skills enhances your ability to sense what's around you.

The human body has been evolving over millennia. From spiritual to intellect, we made gains by gathering and interpreting information from our surroundings and making decisions with the best chances to be successful for our survival. We based these decisions on past experience and learned what was potentially dangerous, what warranted our immediate attention, and what was most likely benign.

The human body has been on Earth longer than any tech gadget, so doesn't it make sense to trust your instincts a little more? You know more than you know. Our bodies are valuable tools for picking up on ghostly things. Don't be swayed by skeptics who say only unbiased machines can prove the existence of spirits!

True, high Electromagnetic Fields, or EMF, can confuse the mind. High levels of either positively- or negatively-charged fields can produce symptoms of paranoia, fear, and that feeling of being watched. This is something to take into account during any investigation. In order to be a credible investigator, you need to eliminate the possibility of contamination due to EMF. I recommend debunking in the moment wherever you can. It's hard to remember when reviewing evidence later exactly what was around you at that time. High tension wires, microwaves, even other equipment or cell phones can give false positives. We want valuable evidence, not mere stories.

This is another reason some folks claim we're just a pseudoscience. Scientists are used to having complete control over a situation and control all variables, but that's not so easy when dealing with spirits, ghosts, and the metaphysical. Much of what we deem paranormal seems to occur irregularly or spontaneously, or at least we haven't discovered what parameters are needed to produce paranormal phenomena. Always more research is needed. Sharing that research and evidence and putting it in a central database would be helpful to advance the field. We rely on what we can collect in the moment and we still have so many questions. Trust your intuition but keep an open yet skeptical mind.

It's pretty likely that it's high EMF when your huge spikes at a location seem to coincide next to the high tension lines near the property. But what about when you just know someone is right behind you, all the hairs are standing up on your neck and you're in an abandoned building in the middle of nowhere? Now this is where experience and training come into play. Young ghost hunters all wound up on adrenaline can easily scare themselves into a frenzy. They psych themselves out and jump at the littlest sound. Every bump is the Boogeyman. Don't scare yourself into an episode of Scooby Doo!

I see this a lot when there are larger groups of people in a haunted location. It's like that old game of telephone.

Someone shuffles their feet. "Did you hear that?" "Was it a voice?" "It sounded demonic!" "This place has evil spirits! Let's get out of here!"

I've seen it happen. Slow down. Much of the time a flickering light is simply a flickering light. That cold spot? I've seen so many people feel a cold breeze and are just sure it was an astral being. When I slowly point towards the air conditioning vent, I've had some folks get angry with me! How dare I ruin their paranormal experience! I assume you want to be a credible paranormal investigator, right? People will take you more seriously in the long run if you investigate a location, did your best to debunk contaminates, didn't find any remarkable evidence, and admit you'll just have to go back another time. A questionable picture of an orb is not going to make you a better investigator.

Being sensitive and being able to differentiate the everyday building noises from something that could possibly be paranormal comes with practice. You could be the best, most psychically aware person on the planet but it's still a learning process.

I trained and showed Morgan horses for over 20 years. Working with 1200-pound animals who don't speak English, you get good at creating a form of communication pretty quickly. Over time, this communication becomes more non-verbal. You get in sync with one another.

It's kind of like that when you go on an investigation. Their thoughts bouncing off you, yours is bouncing off them. You can feel it. You can tell which rooms have a vibe.

Our bodies are one of the best detective tools available to us; we use modern equipment simply to validate and preserve what we already know and detect.

CHAPTER 13 :
What's a Psychic?

So what exactly is a psychic? Are they different than everyone else? Are they born that way, or are these abilities learned? Will they put a spell on your son if he breaks a girl's heart? What do they do during a paranormal investigation? How does a psychic see things, hear things? Can anyone develop these skills?

Every one of us is familiar with the five senses: touch, sight, sound, taste, and smell. From these we delve into the sixth sense, or psychic senses. Some people refer to being able to access this sixth sense as ESP.

ESP stands for Extrasensory Perception. It's the ability to stretch the senses a bit beyond what is considered average or normal. It's not entirely accurate to refer to it as only one additional sense because it's a combination of abilities. It's like having an extension of the normal senses, and these extensions are divided into what are referred to as the Claires.

I'm sure you've all experienced them to some extent at some time or another. Have you entered a room and it feels like people are arguing even though the room is empty? Visited the home of a wealthy Victorian-era lady and smelled old-fashioned perfume?

We all have these inherent abilities. We can all hone and tune these gifts. Some of us are more gifted than others while still others tend to block these abilities. These abilities are gifts and strengths, and everyone embraces them differently. Few people grow up saying they want to be a psychic; it simply becomes a calling and a passion.

We are used to the five senses. These are more tangible, more accessible. When we get into the senses of interpreting what is just beyond what we can barely perceive, it's unnatural in the beginning.

We feel the hair standing up on our arms but don't know how to interpret it. Our ability to perceive something is based on belief and knowledge, or experience. We have to redefine what is normal.

Take, for instance, a fly caught in a spider's web. For the spider, this is not only normal but expected. Web was built, lunch has arrived. To the hapless fly, this is chaos. Unexpected and tragic, the fly is panicking at the situation. Same scene, different perspectives.

Psychics are often stereotyped, often by Hollywood, usually depicted as some scarf-wearing old gypsy woman with lots of bangles and gold earrings. In the more modern version we have Whoopie Goldberg from "Ghost" or the kid from "The Sixth Sense," unable to control the scary visions around him.

That isn't how most of this works or looks. The term psychic covers a broad umbrella of abilities. Generally a psychic is a person who can receive information from an outside source directly into his or her mind. But when we call someone a "psychic," what do we really mean? If we all have psychic abilities, what's different about someone who refers to themselves as psychic?

To me being psychic is accepting that everything is connected on a vibrational and spiritual plane. Regardless of your religious doctrine, I think it's probable that there is much more than our existence here on this planet. Since ancient times mankind has believed and tried to understand what consciousness means and science has yet to unlock the mysteries within the brain. Our thoughts, our past, our memories, our whole cognitive processes are infinitely more than what we know to be proven. When we have an experience, something clicks, like some kind of deja vu.

I believe there are no coincidences, and that everything happens for a reason. It may not seem like it when you get a flat tire on your way to work, but because you were delayed with a flat tire, maybe you avoided a terrible accident. The person you thought would love you forever suddenly leaves, but that opens the door for the right person to come into your life.

Embracing and honing psychic abilities – whether you choose to become a professional advisor or utilize it in paranormal investigations – is a personal choice. If you develop these abilities and simply want to connect with higher energies for personal enrichment, that's fine. We are part of an infinite universe. We are connected to every energy and being in the universe.

Even if you just dip your toes in the spiritual water, you we feel more balanced, more at peace. People spend their whole lives running like hamsters on wheels, trying to balance everything in their lives. They commute to work, shop for groceries, take the kids to soccer practice. They worry about job security, their kid's future, if they will ever find true love because they feel empty and incomplete without someone else's approval. They base their self worth on things money and sweat at the gym can acquire. They fail to live a contented life because they are so concerned with achieving outward things they think will bring them happiness.

I'm not saying everyone should quit their jobs so we can all live on mountaintops and eat twigs. In fact, I've got a house full of bizarre antiques because I find comfort in their history and quirky energy. I'm just saying successful psychics focus *inward*. They gain insight and peace through meditation or prayer.

Spend some time practicing bringing your thoughts inward. Bring your attention back to your intention. Know that all will happen as it should happen. Live your life kindly and with curiosity, and perform tasks with the goal of long-term successes. A psychic resonates with being connected to a higher purpose. It's about being humble, having good intentions, and being diligent. That's really what a psychic's purpose is: to connect with a higher purpose.

You have to work at developing these skills. While many consider psychic abilities as gifts, honing these skills takes years of practice and due diligence. A person who wants to work professionally as a psychic has to be able to receive information and interpret it in a way that is useful for the client. This is why psychics charge for their services. We have bills to pay like everyone else. We have to work at it every day. Practice makes for consistency.

But people are human, and we have strengths and weaknesses. This is why many psychics specialize. They resonate better with certain aspects of love and life better than with others. Some can give you direction on whether or not he is Mr. Right or just Mr. Right Now, while others can focus more on financial or future investment planning. Some psychics are healers and are trained in additional modalities such as Reiki. Working with these psychics in conjunction with Western medicine can help confirm a diagnosis and alleviate discomfort faster and more efficiently.

A psychic's job is to receive and interpret information and steer the client in a direction most likely to be successful. We have good days and bad days. Sometimes the information doesn't make sense at the time, but may at a later date. We interpret the information to best assist the client and situation.

Then there are the psychics who talk to dead people. We refer to these psychics as mediums. To communicate and to see those who have passed, a medium needs to raise his or her vibrational level. The spirits on the other side try to match the medium's vibrational level, meeting in the middle, or at medium level.

Mediums are able to see and hear spirits. A medium can acknowledge spirits, give them comfort, and, if needed, help them cross over. A psychic medium usually will also have other psychic skills. For instance, I am a psychic medium, intuitive healer, clairvoyant, clairaudient, and have clairsentient abilities.

I'm also a Reiki Master. Reiki is a form of healing using your hands to channel and balance the flow of energy. This style of hands-on healing was developed by Mikao Usui around 1914. The word Reiki is actually derived from two words: the Japanese word for Divine or

Universal, "Rei," and the Japanese word for energy, "Ki."

Rather than saying it heals outright, I prefer to describe Reiki as a process that channels universal energy and breaks up blockages in a body's chi, or chakras. I am able to pull negative energies out and restore flow of energy throughout the body, which in turn allows the healing universal energy to freely enter the body. This effectively allows the body to heal itself.

It sounds really involved and complicated but it's a very natural process. I have found it works wonders on emotional and physical traumas. I love being able to pull the negative emotional baggage from someone and watch as tears run down their face when they are able to release all of that junk from their mind and body. Then you open their crown chakra on the top of their head and flood their body with warmth and healing.

Reiki works great on animals as well. Animals don't come in with any expectations, which I think lets them absorb it easier than humans. I do use caution when working with rescue or nervous animals; you don't want to come at them with a glowing beam of energy. You can give Reiki from a distance, which is safer and the animals tend to be more accepting that way.

So how do psychics – and especially mediums – receive information, and how do they make sense of it?

There Are Plenty of Dead People

First, let's be clear about something. Some paranormal groups and locations that advertise being haunted get their panties in a twist when we start talking about crossing people over. My personal experience – and the experiences of respected psychics with whom I have worked – has been that it is free will for a spirit to cross over, and once crossed, they have free will to come back. Some may need help finding their way. There are what we call earthbound ghosts who will not cross for their own reasons. Once crossed, a spirit has free will to return if they are comfortable in the place they once lived. They can come and go.

We will not run out of dead people!

A psychic is not going to leave the Stanley Hotel, wiping their hands, saying, "All cleaned up. No more ghosts." It doesn't work that way.

As long as I'm on my soapbox, let me bend the ears of all the paranormal investigators out there. Taunting and abusing is not a part of being a credible investigation. We as a paranormal community should not relinquish compassion and humanity for the sake of a cool YouTube clip. I cringe when I see folks go into hospitals, sanitariums, and battlefields and start shouting stuff like, "I demand you give me a sign of your presence!" These were places that housed great pain and despair, places where the living knew only fear, sadness and confusion. These folks often suffered from mental and physical anguish. They don't need to be yelled at.

So no more poking with a stick. If a spirit makes the effort to communicate, whether or not they specifically ask for help, it is an investigator's duty to offer solace and assist where they can. Sometimes all the dead want is some acknowledgement to know we are interested in what they have to say. They can teach us so much; sometimes they are at peace there but just don't want to be forgotten.

How We Receive Information

So where does a psychic get information? We divide the ability to receive information into what we call the Claires. Claire means *clear*.

Clairvoyant – Clear seeing

With clairvoyance, a psychic receives information through pictures. Whether it's a still snapshot or moving images, something like an old movie, it comes in bits and pieces. Through practice a psychic can discern if these images are messages and not their own visualizations popping into their head.

We may enter a room and "see" a spirit standing in the corner. During a reading, a psychic may see visions of a farm, an old truck, or see people mingling at someone's wedding. These images give us information not only as to who is there, but what was most important to them.

Spirits may even give us images as if looking through their eyes, like living their life for a brief moment. While most people might think seeing pictures and images and stories is the best way to connect, it's still just part of the story, only pieces of information. We must still gather other information and then interpret all this information into something we can understand and the client can relate to.

Science has discovered how we may be "seeing" psychically. These images are seen using the psychic or third eye. We actually "see" with our brain; ours eyes are merely lenses. Scientists have proven that the pineal gland, located in the center of the brain, may be responsible for clairvoyance. The pineal gland is light sensitive and is constructed very much like our own eyeballs – literally a third eye.

Clairaudience – Clear hearing

A psychic can hear spirits speaking, usually just a few words at a time. By asking questions, sometimes we can glean more specific information or get an identity. Sometimes just a short message comes through and they're gone.

It's not always a two-way conversation. Sometimes they just want to deliver a message or say a few words. It must take a lot of energy to speak from the dead, so I don't hold it against them if it's just a few words. This is one of the best if not the most commonly used skills during a paranormal investigation.

Clairaudience can also mean we could enter an abandoned church and hear phantom music playing. Paranormal investigators often believe they hear workers' voices and pickaxes chipping and toiling away when investigating gold mines.

Clear hearing is basically listening to what the spirits have to say. It is one of, if not the most useful tool during paranormal investigations. Sometimes a spirit may be leery of approaching your MelMeter or touching your K2. A person who died 100 years ago would have no idea what those things are. Sometimes talking is easier and certainly more familiar. Think of it as telepathy, because that's basically what it is. They find it easy and comfortable.

We did an investigation at a residence near railroad tracks. I saw at least one spirit in the home, shy and standing in the corners of the room. We set up all manner of infrared, laser, and digital recorders. We set down to do an EVP session.

"Please come and touch the light so we know you're here," I said. "Touch the RemPod and it will light up for you."

He was having no part of it. I started hearing him talk into my head. I asked him silently to simply join us in the center of the room. Another investigator noticed the laser pattern on my arm become distorted when he came close, giving us validation we were communicating with an intelligent spirit.

"Just touch the light for me," I said.

I hear him say, "I don't know, I think they might blow up."

Obviously I knew they weren't going to blow up, thus providing further validation that this was someone else's voice communicating with me.

Speak out loud and introduce yourself. I speak out loud when communicating with spirit to let the other people in the room know what the spirit is saying to me in my head. It's a little like a phone conversation but it works, and is very, very useful during investigations. Let spirits know why you are there.

Whether communicating psychically or capturing spirit voices through electronic methods, I think spirits find this the most natural way to communicate. This is why I focus mainly on collecting audio evidence during investigations, both psychically and through EVPs. Clairaudience encompasses hearing thoughts of the living as well. Mind reading also falls into this category.

Clairalience – Clear smelling

It sounds odd, but the sense of smell is most closely tied to memory. During readings we often smell Grandma's empanadas. Channeling the Virgin Mother is often accompanied by the heady smell of roses. Smelling cigarette smoke when no one in the house smokes? Maybe dear departed Dad is dropping in for a visit.

One of the best and most specific uses of clairalience involved a phantom perfume. I'm using the term of clairalience and not an olfactory phenomenon because only the psychics in our group picked up on it – it wasn't noticed by everyone in the group. We were on an investigation on the RMS Queen Mary with a group of about 20 people. Out of nowhere, the pungent smell of an old-fashioned perfume appeared in a tight clump. You could walk into it and walk out of it. We were surprised everyone didn't notice it. It was in a ball about 3 feet in diameter. Now sometimes when a person wears perfume, it wafts behind them, but this was free floating. It would appear, and then suddenly be just gone. For over two hours this perfume would suddenly appear and then be gone again. What's interesting is the perfume only appeared in the areas of the ship that were open to guests – we never smelled it near the servants' areas.

Clairsentience – Clear feeling

A human's ability to survive depends on them knowing what is going on around them. Call it getting vibes, putting feelers out, feeling out the room. We all do it. How we utilize this information differs.

Ever go into a room and all the hairs stand up and you go into high alert? Have you entered a bar and you can just tell a fight is about to break out because the energy just "feels" different? Have you been on an investigation and one particular room makes you incredibly sad?

Don't be quick to discount what you're feeling. Sure, high EMF can affect your senses and give you these sensations. Pure adrenaline and anxiety can also get your mind on high alert.

This is where you, the experienced paranormal investigator, come into play. When you suddenly feel this way, check for electrical wiring and even plumbing pipes with an EMF detector. Use either a MelMeter or a tri field meter calibrated for "natural" EMFs. These are specifically calibrated to weed out manmade electromagnetic fields such as radio or microwaves. Maybe double check with a simple K2 along wiring to see if it sets off. Debunk in the moment. If no high EMFs are present, then what you feel may be caused by spirits drawing energy from within the room.

It has been said that when spirits try to manifest, they draw in or feed on ambient energies, batteries, generators, even your energy, and emit EMF. This is a common belief among paranormal investigators as to why EMF spikes during bouts of activity. See how getting in tune with these abilities is useful on a paranormal investigation?

Claircognizance – Clear knowing

Kind of a tricky one, claircognizance is simply knowing something to be true even without previous or supporting knowledge, or logic. You just know. This particular thought rings as true as your closest memory, though this information may have suddenly popped into your head.

Clear knowing, which is cerebral by nature, doesn't have a corresponding physical sensation, so it's a little harder to pinpoint. It's also more difficult to develop, because how do you practice "knowing" more? The other intuitive modalities resonate more with our tangible senses, and we tend to find them more comfortable. We still have to interpret the images we receive, and we may not understand the words we hear psychically. It takes practice to separate our own thoughts, feelings, and sensations from the information coming into us.

Other claires are open to interpretation and we may not always be totally accurate. But with this sense of knowing, it's exactly that. It comes in so strongly there is no room to argue the point. It simply is so.

Think of it in terms of little premonitions. When on an investigation, do you find yourself walking solidly down a dark, creepy hallway in a building you've never been in before as if you "know" exactly where you're going? Do you find yourself picking out an outfit you might not normally wear for this occasion but later find it was the most appropriate? Do you go on a blind date and are just sure this is your soulmate? This modality pops up more in life situations rather than in the paranormal. When you feel it, act on it. Don't berate yourself after the fact saying "I just knew it!"

Clairgustance – Clear tasting

This one in my experience isn't as pleasant as it sounds. I rarely get the taste of chocolate chip cookies in my mouth from spirit. The more pleasant food-oriented experiences usually involve smelling baked goods, as opposed to tasting them.

When we experience a taste psychically, it very often involves getting a sudden taste of iron in your mouth – the taste of blood. Typically when a person has died under traumatic circumstances, such as an accident or murder, I will pick up on their presence through the taste of blood in my mouth. It isn't the most pleasant way of channeling, but it often is a good indication of an accurate connection.

Clairgustance can be achieved through psychometry – the holding of an object either from the scene or an object belonging to the deceased. I can get that familiar taste of blood by visiting the crime scene and then connecting to the victim. Once I have determined they died suddenly and tragically, I can press for information on how. This technique is valuable when working on unsolved murder cases. By experiencing their last moments physically, spirits are often able to relay details of their passing and possibly give you glimpses of who may be responsible.

While I don't trans-channel – which is to allow spirits to fully enter my body to facilitate communication – I do allow them to let me carry and feel their pain and suffering for awhile. It gives them a bit of peace knowing someone cares and took the time to acknowledge them. It's a very personal way of connecting to the departed.

As I mentioned before, psychometry is a useful exercise in developing your claires and discovering which ones resonate best with you. We sometimes hold psychometry classes where everyone brings objects that either have some sort of history or came from an historic place.

Scatter the objects around the table and, one by one, hold each in your hands. Don't tell anyone what you're getting yet, and don't reveal the origin of the object you brought until everyone has had a chance to feel each object. Discuss what you felt. There is no wrong answer; what you felt is what you felt. You'll be surprised how a little rock can give away so much about its origin! I have brought rocks collected from the Roswell UFO crash site, stones from the grave of Billy the Kid, antique jewelry, post-mortem photos, even old toys. Try this little game with your friends – it's a fun way to hone your skills!

CHAPTER 14:
Grounding

After an investigation, it's a good idea to complete a personal cleansing. Sure, these places are dusty, dirty and grimy and we all need a shower after 8 hours in an old building. But I'm talking about personal spiritual cleansing. You need to be clear in mind and body so you can go onto the next location without stuff clinging to you.

If you are cleansing the location, use what you brought for that to clear yourself too. Sage everyone who is there with you. Line them up, go up, go down, and don't forget the bottoms of shoes, too.

Mentally clear yourself. Is there any chattering in your head? Did you bring angels in to assist those who needed help? Good. Now wipe the air over your head and clear your mind. Ground yourself.

Repeat this a couple of times. Head to head, chest to chest, hands to hands, butt to butt, legs to legs, feet to feet. It seems simple but it's all about getting your mind in the right place. You're in body. You're in charge. No hitch hikers. You don't want spirits or any forms of energies clinging to you.

Some people think it would be fun to have a ghost follow them home, but the real world is not the same as an attraction at Disneyland. Think about it. Why would you want an energy attached to you, draining your energy? That's what they'll do, and feeling drained and tired is the first step toward being vulnerable – and being vulnerable is not a good thing in this field.

Some people get a form of the munchies after an investigation. Eat something high in protein. It's great for grounding. Then go home and take that hot shower. You've been investigating all night.

I usually need a Mountain Dew.

CHAPTER 15:
Boundaries and Protection

Some psychics – and even those who are just extra sensitive – sometimes have trouble filtering all the information around them. They feel constantly bombarded and drained by all the energies.

Boundaries are essential. For someone who works with another's energy, whether living or in spirit, you must have control over how much you let in. Boundaries are an integral part of protection. Spirit can attach to people like Velcro. By the end of the day you might feel as though you're covered in sticky notes.

Personal boundaries from a psychic or paranormal investigation standpoint are very similar to personal space. It's a matter of respect. You can't interpret any information if you allow it to pelt you from all sides. Remember, you are allowing whatever it is to come in. You have free will. You are the one who has the physical body, not them. You know that old folklore that a vampire can't come into your home unless you allow it to? Well, it works just like that.

I know some psychics who trans channel, which is letting a spirit enter their physical body in order to communicate. I do not trans channel. I ask spirit to stand next to me. I can still feel their pain and emotions, but I also have the control to tell them to take a step back if they become too overbearing. Spirits and ghosts can be deceiving. They might convince you they are some sweet old lady but they hop inside you and it's a whole different story! Sometimes the spirit may be friendly but really likes the feeling of a human body and could just want to stay for a while – a long while.

Boundaries are equally important for people living in active locations. These folks are living with an energy that is not in their dimension. These spirits may be friendly and protecting, like a sweet Auntie Em, puttering in the kitchen baking cookies and watching over the children. (And if she would just do all the house cleaning while we are at work, I think that would be every paranormal investigator's dream!) However, we have to remember that spirits were once people, and there are nice people and some who are ogling peeping toms. They don't change their personalities much on the other side.

I recently had a case at a private residence near railroad tracks. Imagine the hard working laborers toiling in dangerous conditions, just trying to better their lives. But the kinetic nature of a railroad lends itself to attracting a seedier type of worker. I think there was one of each in this young woman's home.

She had reports of shadows darting back and forth. A large shadow form would stand in her bedroom doorway, and it eventually progressed to an entity sitting on the edge of her bed, sinking the bed corner down with its weight. Her young daughter was having conversations with invisible friends. She really became alarmed when her mother slept over and experienced what may have been a succubus, or night terror. The grandmother found herself pinned to the bed with a heavy weight on her chest, unable to breathe, unable to see what held her. The family was rightfully upset and asked for our help.

During our pre-investigation interview, we noticed shadows dashing passed the mother. We heard disembodied voices coming from another room. In the middle of the interview, one of our female investigators suddenly became violently nauseous and sick to her stomach, and had to rush out of the room. This was before we even started our investigation.

The mother, grandmother, and young daughter left the house for the night so we could have the house to ourselves. We blanketed the house with lasers and infrared in the bathroom, where the reports of a spirit sitting on the bed and shadows in the doorway had originated. We measured high EMFs in some of the electrical switches. We observed that the house was situated near high-tension wires, which can cause high ambient EMFs and affect cognitive abilities. We had that place canvassed, ready to capture this aggressive spirit.

We were able to make contact with a shy, stuttering spirit who I believe worked and died on the railroad. I was picking up the name Danny. Danny found it easier and more comfortable speaking through me rather than manipulate our electronics. When we asked him to come and touch something so we knew he could hear us, I heard him say, "I don't know, I think they might blow up!" He must have thought they were dynamite, which was something frequently used in railroad construction and a material that led to the deaths of many a railroad worker. Perhaps this Danny died from a dynamite explosion.

We were able to have Danny come close enough to disrupt the laser grid on my arm, indicating he came into the room. He also gave us that pins-and-needles feeling in our arms, probably using our energy to communicate with us. What was interesting was this Danny was clearly the meeker of the two spirits in the home. He even indicated to us that he was bullied by the other spirit. That is something I had encountered before. A "bully" spirit will taunt and abuse other more submissive spirits. I think both spirits initially came in because the high tension wires gave them energy to do so, but also the females living in the household had the kind of energy that would attract lonely railroad workers – even dead ones.

This created a problem. During our investigation, we were only able to connect with Danny. Where was this bully spirit? We recorded no shadow activity, no K2 hits, no other laser disturbances. Where had he gone? It seemed Danny felt comfortable approaching us that night because he had the house to himself. You would have thought a more assertive spirit would want themselves known.

Sure enough, within days of our investigation, the homeowner told us she experienced activity where she stayed while we were in her house investigating. The aggressive spirit had simply followed her for a slumber party and returned to his usual mayhem and abuse when everyone returned home.

I had a chance to sit down with the client before our investigation. The homeowner is an important dynamic of a paranormally active location. I like to get a feel where they're at energetically, because they affect, consciously or subconsciously, the energy within their home. This homeowner had very unbalanced chakras. Being a nurse, her throat and heart chakras were very strong; she was someone who regularly had to communicate in emergency situations and had a giving nature. But all her self-protective chakras – everything that helps her be safe and balanced – were some of the weakest I've ever encountered. This is where your sensitive and psychic training are important. We now knew this woman lacked the tools to really protect herself from energy attacks. This is when an aggressive spirit can become dangerous. They can pick up on this weakness and take advantage of it.

The spirit in her house was not only clever, but feeling very entitled. That's a dangerous combination. An energy that feels entitled to your space can and will wreak havoc. They will wreak havoc on your environment but also may manipulate your thoughts and moods. This makes a living person who is already vulnerable even more susceptible to negative attacks, and this can lead to possession. There is an argument that many cases of mental instability and insanity is caused through external forces. Something energetically outside the person's control is manipulating their thoughts and, ultimately, their actions.

The paranormal activity continued in this woman's home. We offered to return but she refused, saying she could handle it herself. Could she? Should she? As of this writing she is not willing to have us return, yet she is still frightened in her own home.

Establishing boundaries is paramount for your own safety. It's a matter of respect and personal space. If some dirty, ogling, pervert came into your home and watched you taking a bath, would that be OK? Probably not. So why should it be OK with a spirit having that same mindset? Whether it's because you want to experience all aspects of the paranormal, or you think it's cool and think you can take it, or perhaps you just are in denial and want to avoid a confrontation and think it will all just go away, leaving yourself open is just not a good idea.

Be aware of what's out there, and know how to stay safe, energetically and physically.

Setting boundaries is especially important when investigating hostile locations. Setting your intentions before entering such a place let's them know upfront that you're not going to tolerate any funny business from them. Always set this boundary intention, but be extra assertive when going into prisons or wherever aggressive energy has been found. Stick your chest out and put your head up in the air. Be open to what's there but wary of lower vibrational energies that try to sneak in. They will try. Anything of a lower vibration is opportunistic and will be attracted to energy and your light. Keep them at a distance. They drain your energy and affect your mood. Neither is productive or safe.

But how do you protect yourself? Sometimes it's as simple as mentally setting your intention that you are in control and you work in the light. Your intention is for the highest good. If you prefer more ritual, many wear a St Benedict medal and say a prayer to St. Benedict before investigations. Often I ask St. Michael to guide me. You can always ask to send angels ahead to a location to hold sacred space.

Repeating the following phrase often gets you in the right frame of mind:

I am safe, happy and whole. I come with respect and ask questions for the highest good. I work and live in the light. I am part of the greater collective. I am part of the universe and have the strength of golden light. Wherever I am, God is.

You can personalize it, of course. Whatever resonates with you is best, but you get the general idea. Having all members of the investigation join in a ritualistic protection is a good idea. Some folks have different religious views. Just go with working in the Light and come with respect. Spirit will get the point. It's all about intention and owning your space.

I don't recommend letting spirits feed off your energy unless you are absolutely sure about their intentions and have the experience to gauge their energy. It's best to have them stay next to you and connect. If you are unsure about a location, send angels ahead. When you enter a location that feels off, imagine sending golden white light ahead of you and all around you. You are made of the eternal universe, and have a higher purpose. Remember that.

CHAPTER 16:

Putting It Together

We get all of this information through many senses. Most psychics have one or two strong points. If they are strong in clairvoyance, or clear seeing, then they might be able to get information by receiving pictures of the past. They might see them walking in clothing from the period, or see a woman sitting in a chair. We may see only a single item, such as a miner's pick, and have to interpret this information. We ask for more.

If the psychic's strength is clairaudience, they might hear music playing or hear a child's laughter. I'm good at picking up feelings for a room or an individual, which is clairsentience. I can feel their emotions. I also get word or thought messages, a sense of knowing. And I hate to say it but sometimes the information doesn't make sense at all. We'll have good days and we'll have days when we can't tell what they are trying to tell us. It takes patience and practice.

I wish we could just go into a haunted location and get all the information we need in a clear-cut format. "I'm Anna Corbin, here's the name of the person who killed me, thanks for helping me solve my murder." Wouldn't that be great? One day I hope to have that kind of communication, but it doesn't work that way.

Psychics tend to receive information in bits and pieces and we have to interpret these pieces. Is what you feel yours or theirs? We feel a female presence. We ask her name. We may see the letter A. We ask "Why are you here?" We might taste blood. Then we get a vision of a woman in a bathtub, bloody. As we press for answers, we may feel overbearing sadness – her sadness, her panic. Once again we have to question, is it our emotion or are we feeling their pain and fear?

We use this interaction to tell them it's only a memory. Calm them down, bring in angels. Tell them we're here to help. (That is what we're here for, right?) If we have equipment running, this is where we are likely to get EVPs during this conversation. This is how you enhance and direct a paranormal investigation to get the best results. If they are trying to communicate with you, ask them to speak out loud or try to get the spirit to come close to your EMF equipment to document their presence on camera. Start the conversation and see where it leads. All those blinking boxes may look intimidating to someone who died in the 1800s, so go easy. Trust what you feel and always ask them to speak loudly and slowly, because sometimes the feelings and information comes in so fast. Sometimes they've waited a long time for someone to hear them.

If you are in a location where the deceased were emotionally or mentally challenged, go easy and be extra calming. We certainly don't want to traumatize them by demanding they touch scary blinking boxes or push objects off of tables for our amusement. Think in terms of how a counselor would treat a living person in such a situation. Someone has not been able to communicate for years, maybe decades or centuries. Alone and scared, their voices have not been heard. People walk back and forth, going about their business, seemingly ignoring them. Suddenly you guys come along, and can see them, perhaps, and want to talk to them. It could be as jarring for them as it is for us.

I don't know how they manage to make themselves heard, or how they create disturbances in EMF to create EVPs. They may still be trying to figure it out themselves. Try a few different pieces of equipment. If an Ovilus is too scary or complicated, try a K2 or partially unscrewing a flashlight if making it light is easier for them. Remember the spirit who told me, "I don't know, I think they might blow up," when asked to come close to a RemPod? That was one of those moments when just talking seemed to be the most natural form of communication for both sides. Just talk. I mean, you're still conversing with the dead, which is pretty amazing in and of itself. Come back another time with equipment and maybe they will feel more apt to try one of your fancy electronic devices.

As a psychic, consider yourself a translator as well as another piece of equipment. You can be the bridge that spirit needs to communicate. Don't discount what you feel in a location. Sometimes these spirits have been alone for so long that any information comes in too quickly. They all start talking at once. Tell them to speak one at a time. Tell them to slow down. Are you in a location where people who have passed there may be emotionally or mentally challenged? The energy will have a nervous, eerie feeling. Have extra patience with them. These spirits can sometimes be a little touchy-feely and not so good with personal space. Don't hold it against them. You're there to help, remember?

Do you feel bullied or want to run out of the space? Do some spaces make you feel like everything is closing in and you feel as though you're physically being pushed? Stop immediately. Take a deep breath. Feel what's around you. Are the spirits negative lower energies, or just annoying bullies? Are you in a location where the deceased may have been emotionally or mentally challenged and not the best at giving personal space? Either way, take a moment and stand your ground. You mustn't allow yourself to be bullied. Remember to set boundaries and stick to them.

An example of this happened to me when I investigated the Graber Olive house in Ontario, California. The 100-year-old family-run business started with farming and progressed to olive production. This was not exactly what you would consider evil farm incarnate. Nevertheless, Graber Olive house is known to have what's called a creeper. A creeper is basically a lower-vibrational entity, also known as elementals.

We were investigating with EVP specialists Mark and Debby Constantino and we were up in the loft. It was about 2 a.m. and Debbie was getting some interesting EVPs, doing short burst sessions, listening back and having the next line of questions reflect what we heard from the responses. Sitting in a dark room, it's difficult to keep your eyes focused, so I let my eyes wander around. Up on the ceiling of the loft was a creature I had never seen before, but I knew instantly it was the creeper. It was about 10 feet long, humanoid in shape, but it had long, sinewy arms that were also about 8 feet long. The eeriest part was that it was crawling along the ceiling, tilting its head backwards in a very disturbing manner, watching us. When it moved, its feet made a *thwp, thwp, thwp* sucking noise. Why was no one else seeing this? It was icky. Unfortunately it did not show up in photographs, but it was an experience I will never forget.

After an encounter like that, I thought the rest of the evening would be pretty benign – but I was wrong. We went into a room that was used for packing and labeling the canned olives. It was a pretty average room. I walked to the very back of the room and turned to face the center of the room. Several other people were there, including Mark and Debby, a news reporter, psychic Kimberly Rinaldi, and an investigator who had also brought his mother with him along on the investigation. The investigator and his mother had recently returned from an investigation at the Shanghai Tunnels. His mother had been attacked by some negative energy there to the point of her having seizures. We knew we had to keep a wary eye on his mother here. I immediately felt like I was being pushed from behind. I didn't provoke, but I stood my ground. I was as passive as possible without being bullied.

The news reporter was the first to feel oppressed. She suddenly began feeling ill and just plopped on the floor. I began feeling twinges in my gut too. I told the spirits we wouldn't be there long but we needed to stay for a little while. The pressure would recede for a bit. The reporter began rocking back and forth and became quite nauseous. She had had enough and left the room. Kimberly said she felt pressure too. I was feeling poked again on my back. This entity was really becoming more and more aggressive.

I found a little trick that works with this type of energy. It is used a lot in dealing with aggressive dogs. It's a little like passive-aggressive behavior. Without turning to face the energy, you can back up into it.

It's less confrontational if you keep your back to it. Back it up out of your personal space. Then let it have its space and you can step back into the room. If it pokes you again, back up to it again, this time a little further. You don't want it to leave, but you are establishing boundaries. It is a form of respect. It's a form of communication. I acknowledged it, let it have its space, but wasn't going to let it make me feel vulnerable. Kimberly stood her ground, but had also had enough and left the room. I didn't want to aggravate this spirit but let it know I was leaving on my terms, and left the room.

So then the reporter, Kimberly and I were sitting in the hall getting some air and wondering why the other people would want to stay in there. It didn't want to talk and wanted to be left alone. Suddenly, the investigator came rushing out escorting his mom, who looked pale and was having difficulty walking. Her actions resembled the seizures she experienced in the Shanghai Tunnels. Had it attacked her? I believe it did. We had to get her completely off the property for a while before her symptoms returned to normal. I believe negative entities will attack the vulnerable. I think they may more aggressively attack if feeling cornered.

Keep an eye on all parties in your group. If someone is feeling vulnerable, point out alternatives to leaving. Breathe. Establish boundaries. Back into the energy to claim space, but don't corner it. Tell it you're leaving when you leave – don't just walk out.

Speaking of energies, remember the daytime visit to the Titanic exhibit? We were experiencing emotions attached to the personal objects. The jewelry and grooming items that had personal attachment to those lost in the tragedy were still filled with their energies, and we felt them. If you sense emotions suddenly, gauge whether or not it is your own by moving in and out of it. Is it sudden? Does it dissipate when you step away? Try to picture the owner of the objects. Ask them to step forward. Just start talking to them as if they were standing next to you, because chances are they will be.

Objects that are associated with tragedy can have a strong hold on the deceased. The Titanic exhibit had several pieces recovered from the boiler rooms. Eyewitness accounts tell us many mechanics volunteered to stay behind to keep the ship afloat as long as possible. Their selfless acts probably saved many lives. When I entered the boiler room exhibit, I started crying again. This time it was different though. I kept feeling their resolution to remain at all costs. Their dedication to stay at their posts remained long after death. We picked up on one main spirit who was very strong. He still seemed to be working to save the long sunken ship. He wouldn't turn around, keeping his head down and working at some phantom machinery. This is when you bring in help. This is when you call in angels. It doesn't matter what religion you are, just ask for someone from a higher plane to assist you. Don't leave until you see a divine hand on the deceased's shoulder. Let them do the heavy work.

You are there to bring in guidance. Trust me on this, it works. Bringing in angels to help ghosts crossover will not make a haunted location devoid of activity. If anything, it lets spirits know we are here to communicate, help, and ultimately learn from them. It actually attracts more spirits to an area! When you let them know your intentions, I think they tell others, and a location can become a portal of sorts.

Remember the ethics of connecting to those who have died, especially if it was under tragic circumstances. If we, as paranormal investigators, go into a place and seek out voices from the other side, then we better be willing and able to do something when they answer. We don't have to have to have all the answers, but I think having a positive intention goes a long way.

RMS Queen Mary Long Beach, Ca

Encase all the memories and imprints of all the souls who walked her decks. Water, iron, metal, all tightly holding onto those thousands of memories and lost souls...It's no wonder the Queen Mary is considered one of the most haunted locations in the country.

CHAPTER 17:
RMS Queen Mary Investigation

One of my favorite haunts is the beautiful RMS Queen Mary. Setting sail in 1936, she is an Art Deco masterpiece. It is so beautiful. I really get a sense of stepping back in time aboard the ship. Walking her decks is inviting, whether or not you get any evidence. It's forming this bond with locations that I think gets you more and better evidence. Spirit can feel it.

The Queen Mary was the fastest ship in the world in her heyday. Called into military duty in 1941, the Queen Mary served as a troop transport, able to carry over 15,000 troops across the Atlantic. She was such a valuable asset to the war effort; she carried a bounty to any enemy sub who could sink her. Painted gray to camouflage her, she was dubbed the Gray Ghost. Indeed, she was such an asset in transporting troops quickly that Prime Minister Winston Churchill credited the Queen Mary for shortening the war by as much as a year. During her five-year war service she carried over 800,000 troops. With this many troops to protect, she had to go straight and fast. While crossing the Atlantic, she was accompanied by several smaller escort ships. These escorts would criss-cross her trajectory to search for enemy submarines.

On Oct 2, 1942, one of her escort ships, the Curacao, tragically crossed directly in her path. The Queen Mary, traveling at over 25 knots, sliced through the tiny Curacao. Carrying over 20.000 troops, the Queen Mary was under strict orders to continue on, and under no circumstances to stop and come to the aid of the crew of her escort ship. If she did, she would have jeopardized all her troops. Over 300 were sent into the frigid water, and 239 of those lives were lost. Some say that's why the Queen Mary is so haunted. Her wartime service surely stained her forever and left pockets of despair and sadness within her. Not only did she carry troops across the Atlantic to fight, she transported prisoners of war to be relocated in prison camps. Contained in the poorly ventilated cargo holds, many captured soldiers succumbed under the stifling conditions. Most were simply buried at sea, never to make it home.

Maybe it's the combination of Art Deco regalia and years of happy passengers traveling to new lands combined with the memories of thousands of troops who might not have returned. Encase all this energy, those memories, the imprints of all the souls who walked her decks. Then contain all those memories and energies into a steel ship. Water, iron, metal, all tightly holding onto thousands of memories and lost souls.. It's no wonder the Queen Mary is considered one of the most haunted locations in the country..

Because I live so close to the Queen Mary I sometimes visit the grand ship on a day off from work. I think I can justify it by saying climbing up and down all those stairs for hours is way better than going to the gym. But let's face it, I just love the sight of her, her artwork, her murals, her exotic woodwork, her original 6 foot golden alabaster lights in the Queen's Salon, Oh. and her many, many ghosts..

I have so many personal experiences aboard the Queen Mary I have lost count. Years of exploring and investigating have captured loads of evidence, even during a daytime wandering. Every area of the ship is loaded with energy. Some areas of the ship do make me feel a sudden feeling of extreme sadness and despair. The vast majority of the ship still has a sense of excitement and curiosity. I believe the ship has many spirits who loved the ship as much as I do and had happy memories here. Here are some highlights of my many travels aboard the Queen Mary..

Once, a friend and I were roaming the halls on B Deck in the late morning. Suddenly a woman appeared, having quickly turned a corner, and began walking alongside us. She was dressed in period 40's clothing complete with short, white gloves. I saw her and thought they must be hiring people to work in character on the ship. The QM holds a Roaring 20′s Gala every year, so I didn't think it too out of place. Except the Gala was months ago, and no one else seemed to be dressed up. Then, the woman, still walking right next to us, began talking. She was saying things that sounded a little... odd.

"I just love this ship. I just stay here," she said.

"Yes, It's a beautiful ship," I responded.

She walked alongside of us and as she did she brushed the veneered hallway with her gloved hand. Her fingertips just lightly brushing the walls as she walked with us. OK, a little weird, I thought, but I figured she was just staying in character. She looked absolutely solid and real but there was just something so strange about her.

Then without saying Good Morning, she turned a corner and was just gone. My friend and I looked at each other puzzled. Was she... real? I ran back to where she had just turned and she was nowhere to be seen. She couldn't have entered a room so quickly and we heard no doors opening. She was just..gone..Gone as quickly she appeared. This was at 10 in the morning folks..

One of the smallest and yet most active spots on the ship is the Isolation Ward. Barely two rooms, it remains intact with original 1930's sinks and wooden bunk beds for patients. Simple surgeries were even performed in this area. Plaques hang on the walls listing the names of passengers and crew who died while aboard. A few were accidents (such as the crushing death of seaman John Pedder in watertight door 13) but most were from natural causes. The small room that still has the old patients' beds always seems to have a sadness about it. Investigators often feel ill in this area. Once, another investigator, Patty, and I were doing a quick EVP session in this area. Suddenly, I felt a sharp pain in my stomach. I looked at Patty and noticed she too, was holding her stomach. At first we thought it might have been some evil in our lunch we had earlier. We stepped out of the area and the feelings immediately dissipated. We looked at each other a little surprised. We stepped back into the room and the nausea returned. Unpleasant as it may feel, this is connecting with the energy left behind by spirit. If strong enough, it may be a way for an intelligent spirit to try to communicate. Your empathic abilities will pick up on how they felt when they were a patient here. Notice how it dissipated immediately when we stepped out? That's a good sign it is an energetic influence confined to that area and not just an upset tummy. Trust your gut. (pun intended).

The other side of the Infirmary has a completely different feel even though they are only 20 feet apart. This side has that excited feeling that permeates throughout the ship and it's in this room we have captured some of our best Class A EVPs. One of our clearest was captured before we even asked who was with us. It says, " Rosemary...Is my name." It's so clear people who first heard it thought it was another investigator in the room with us. We have also captured a kitten's "meow" in EVPs in this area. Along with great EVPs, we have also heard audible singing and conversations in this tiny room.

The Boiler Room is another favorite spot below decks. The boilers removed, it's now a cavernous space with wooden catwalks above the iron girders. This is an area where I once investigated with Jason Hawes and Grant Wilson. All was quiet and then a huge crash of glass smashing as if someone threw a heavy glass wine bottle against the wall. We all ran over to the area but there was no glass, no bottle.

The Boiler Room is also the area where I succumbed to temptation and yes, I used a Ouija board! I'm in a dark, cavernous, iron room of a haunted ship at 3 AM with a Ouija board and author Jeff Belanger saying "Wouldn't this be fun?" Yes, I succumbed to temptation. Sure enough, the planchette danced under our fingertips. As someone held a flashlight overhead we asked. "How many of us are here?" Our eyes widened as it counted each of us. It kept pulling us down to the months of the year at the bottom of the board. It circled a bit and landed on September. It then zoomed up to the number 2 followed by the number 6. It then grew quiet and we wondered what it meant.

The next morning I was walking through the exhibits with a girl who was with me during the Ouija board session. Suddenly, she grabs my arm and points excitedly to a sign. "That's the date, That's the date!" Sept 26, 1934. The date this ship was christened the Queen Mary! I love this ship!

The Engine Room has a completely different vibe. Here, polished brass gauges and original turbines sit quietly and seemingly idle until the calm of night falls and the visitors are gone. Then it becomes alive with the voices and shadows of the past. Here it is best to simply sit quietly and enjoy the spirits come to life in this room that was once the heart of this massive ship. Some of the activity seems residual - merely bits of history being replayed over and over, oblivious to either us or time. We hear the huge brass spigots squeak and turn. We see shadows of men against the walls, darting quickly back and forth. The infamous watertight door 13, where John Pedder was crushed, is in "Shaft Alley' here. Shadows dart in the corners here and sometimes, just sometimes, when people walk through the watertight doorway, they find handprints of oil on them. The engine room is also a place where we've heard audible phenomena. We'll hear voices and bits of conversations. This has to be someone else in the engine room with us, we think. We check. Nope. Not living, anyway.

Sometimes we can acheive a little interaction by striking the railings with a bar. We often get clangs in response, coming from the far end of the engine room. This was a common form of communication when the engine room was operational as it was extremely noisy with all the turbines.

The most interesting interaction I captured in the engine room was when psychic Chris Fleming and I were utilizing an early form of the Ovilus. Created by Digital Dowsing's Bill Chappell, it contains a word database that can be utilized by manipulating the EMF around it. each word corresponds to a particular fluctuation in EMF. The words "hot" and "tired" were repeated several times. The engine room would soar well over 100 degrees when in operation. That would be considered quite “hot”. I tried to coax a bit more attitude from these reclusive sailors.

I asked, “Do you know there are a bunch of women in this engine room?”

The word “Bitch” blurted out from the Ovilus. This early version did not have profanity programmed in and it did not yet have a phonetic mode. How did they do that?

“You did NOT just say that!” I scoffed.

“Sorry” it huffed.

Evidence can occur in the hotel section and during the daylight hours as well. I was with psychic Kimberly Rinaldi, a good friend of mine, and we decided to take their afternoon tour and putter around. We were in a group of about 10 people, plus our guide. She took us to a room where a full-bodied apparition had been seen. While looking around and taking photos, we suddenly were hit with a very strong women’s perfume. Just, Wham! in your face and strong. We figured someone was wearing a lot of cologne. This was a flowery, old fashioned perfume. We moved on to another area, and there it was again. Wham! You could step in the bubble of scent, and step out. It wasn’t wafting where someone had been walking, it was manifesting in strong clumps and appeared suddenly and then vanished suddenly. OK, we thought, it was time to debunk this. We proceeded to sneak behind each guest and sniff each one. Standing behind a guest. Sniff. Nope. This one? Sniff sniff. Nope. Act casual, now. Sniff, sniff. (Who says ghost hunting can’t be funny?) This scent followed us for over two hours, and what was even more unusual was that the perfume only manifested in guest areas and not service areas.

I had mentioned this story when discussing clairgustance because we seemed to be the only ones able to smell this strong, strong perfume. I can’t say definitively it was an example of clear or psychic scent, or just a phantom perfume. But it was clearly not coming from a living source.

It seems anywhere you walk on the Queen Mary you might find something paranormal, day or night. The rooms are active, with water turning on, or lights switching on and off by themselves. You may see an apparition wandering the halls. We have seen transparent spirits in vintage clothes leaning over the railing on the Promenade deck, looking out into the ocean. But the most active and most popular of all is the Pool Room.

The Queen Mary 1st Class Pool Room

Does a little girl's spirit haunt this room?

Photo courtesy of the RMS Queen Mary archives

CHAPTER 18:
RMS Queen Mary Pool Room

The beautiful Art Deco masterpiece that is the Queen Mary has many active haunts. The most popular for its paranormal activity is the first class pool room.

The first class pool room is festooned with a Mother of Pearl coved ceiling, diving boards, a balcony, and changing rooms. The pool room was filled with salt water and heated by the ship's boilers. During tea, the second class passengers were allowed to swim in the fancier pool whilst first class passengers enjoyed their tea. After tea, the pool was drained and refilled before the first class passengers returned. During wartime, the pool was drained and used as extra sleeping quarters. Space being at a premium, bunks were stacked three-high to utilize every inch of space.

Some say the changing rooms contain a vortex. A vortex is a thinning of the veil between worlds, thus allowing spirits to cross easily into ours. Feelings of lightheadedness, disorientation, and vertigo are often felt in the changing area.

To me, one of the creepiest areas in the pool room is the space underneath the stairs. The story goes that a man raped and murdered a woman and stuffed her body underneath the stairs. I have not been able to find documentation of this murder, but the energy is very negative in this one place. While investigating with a group, I sent a few people back into the space. I gave no details, only told them to feel the difference in the energy. One man returned, white as sheet. I'll never forget what he said.

"I sure hope nobody ever took a woman back there."

Then I told him the story. Then he really got pale!

But the pool room is most noted for what is believed to be the spirit of a young girl, Jackie. Jackie supposedly drowned in the second class pool, but no record exists of such a drowning. This is a case where I think a spirit has taken on a role it enjoys. Jackie loves interacting with people and while I am usually skeptical of child spirits, I have never felt anything malevolent from Jackie.

The famous psychic Peter James had many documented interactions with Jackie. They seemed to have a special bond. There is video evidence of "Hellos" and "London Bridge is falling down," and laughter between Peter and Jackie.

Peter James loved the Queen Mary so much that when he passed, his cremains were aboard the ship for some time.

Jackie and I have had lengthy K2 sessions and interactions. She will answer any number of questions, but when asked if she is a 6-year-old child, the K2 stalls and she refuses to answer yes. An honest ghost!

One of the endearing aspects of Jackie is that not only will she give you audible responses, she will let you know when she's tired. When she gets ready to leave, you can feel a cool breeze wrap around you and then pass onto the next person. They say that's how Jackie gives hugs.

While taking shots using an infrared still camera, I was capturing pictures of a psychic who was allowing spirits to draw energy from her. She complained of feeling drained and in pain. (Remember what I said about boundaries?) We were in almost complete darkness with just a few K2 lights. When I looked at the photos I took of her, everything was black except her hands and arms, which glowed purple. No lighting of any kind was on her, but so much energy was being exchanged that it actually glowed in infrared.

Along with Jackie, several other spirits have been experienced in the pool room. Since the pool room is known for audible phenomena, we thought we would try an experiment. During an investigation, we thought we would try to reach some of the many soldiers that had been aboard the QM. A member of our group had spent many years in the Navy. We decided to use him as a trigger object.

“Can you sing our song to inspire us?” he asked. “Can you sing our song, ‘Anchor’s Away’?”

We listened. Slowly, from the balcony, came not one, not three, but what sounded like a chorus of 30 spirit voices rising up and singing “Anchors Away.” I still get the chills!

The drawback of audible voice phenomena is that it seldom records, unlike an EVP that is recorded in the white noise. So while it is a great memory to me, it is, scientifically speaking, just a story.

CHAPTER 19: Psychic's View of the Queen Mary Pool Room

The Queen Mary is a gold mine for collecting evidence with every blinking box I have. It's beautiful to photograph from any angle. Using all your senses and psychic abilities, let's go on a trip aboard the Queen Mary and see how a psychic contributes to a scientific paranormal investigation.

First thing's first: ask for protection and guidance before any investigation. No matter how friendly the spirits are at a location, something can always be hanging around. They see you as a bright light and may seek you out. When you open yourself psychically, by design it makes you somewhat vulnerable. Get a feel for the environment before you board the ship. Are you compelled to look at certain areas? Getting any weird vibes? Is anyone talking to you in your head? Don't worry if you feel nothing. Just allow and look around.

If you booked a room, does the air feel different inside? Is it heavy? Be aware of this type of change in each area. It can indicate a change in barometric pressure, which coincides with paranormal activity.

How do you feel in the isolation ward? Where there was sickness and especially nausea from seasickness, as a sensitive you may feel nauseas yourself. Just take a moment and think. Are you actually sick, or are you feeling someone else's illness? Step out of the room; does it dissipate? Step back in the room; does it return? If you are picking up on this, tell other investigators, as they might be feeling the queasiness too. Now would be a good time for an EVP session. Spirit is present. Let them know you feel their discomfort and are here to be of assistance.

You may not pick up much in the engine room as it is mostly residual energies. They don't often interact, so don't be discouraged. Touch some of the instruments and gauges. Do any of them have a vibration that differs from the others?

A good place to go from the engine room is watertight door 13. Door 13 is where 19-year-old sailor John Pedder was found crushed by the heavy automatically closing door. It is not known for sure how he became wedged but it was believed he was playing a game of "chicken" during a drill, darting back and forth through the doorway. I have experienced tight balls of energy in the walkway on either side of the door. On an investigation with psychic Chris Fleming, Chris stopped in his tracks in the walkway, having literally "bumped" into one of those balls of energy. It's difficult to get any audio evidence in this area because it's right by an elevator shaft and near a noisy water heater. Feel what you can and have investigators take EMF readings and measure temperature changes.

The pool room is where your gifts can truly shine. Most investigators are aware that this area has the most intelligent spirit activity on the ship. Famous psychic Peter James spent many trips here having audible conversations with the spirit known as Jackie. Remember her? She is supposedly a little girl who drowned on the ship, but I have not seen records of this death and I tend to be cautious of child spirits. Personally, I think this Jackie is a bit of a showboating spirit. She craves attention and interacts vocally, with K2s, and dims lights. She's a lot of fun. Try singing to her and see if you can get her to respond.

In recent years I have encountered more soldier and sailor spirits in the pool room. Remember the story of the voices singing "Anchors Away"? Be open to what might be there and see how much interaction you can get. Have your team record audio and video.

The only real sad spot in the pool room is underneath the stairs. As I mentioned earlier, this area is filled with energy, possibly from a woman who was murdered there. Queen Mary officials won't confirm or deny, or elaborate on the story, so it's hard to find the truth. But go under the stairs and feel the difference in the energy. Feel the sadness.

Most of all, just be aware of what's around you.

CHAPTER 20:
The Stanley Hotel

The Stanley Hotel in Estes Park, Co

Author Stephen King wrote the Shining based on experiences he had at the Stanley. Would YOU be afraid to stay the night?

"Heeeere's Johnny!"

We've all seen Stanley Kubrick's "The Shining." But have you ever wondered what on earth gave Stephen King the creepy inspiration to terrorize us and make us afraid to enter hedge mazes?

Enter the Stanley Hotel in beautiful Estes Park, Colorado.

This nearly deserted resort is where Stephen King stayed – and was inspired. It was at the Stanley Hotel that the writer encountered many of the paranormal events that made their way into the book. Despite checking in to Room 217, many guests have been unable to actually stay the night in that room, including actor Jim Carrey, who supposedly ran from the room within hours after arriving. Then there's the infamous visit during the second season of the reality TV show "Ghost Hunters." They were in Room 401, Lord Dunraven's room. With cameras fixed on the closet door and others facing Jason's bed, we see the closet swing open, and Jason's drinking glass crack and smash to pieces on film.

The Stanley Hotel is at the top of every list of America's most haunted locations. What makes it so active and, what's more, so *consistently* active?

Built in 1909 by F.O. Stanley, it was meant to be a summer retreat to ease the tuberculosis symptoms of Stanley and his wife, Flora. They bought the property from Lord Dunraven after a hunting trip in the area. The resort consists of several original buildings, the main hotel, music hall, and carriage house.

After its initial purpose to house buggies subsided, the carriage house was used for basic storage. What's interesting about this is when a guest passed away at the Stanley, the mattress they had slept on was stored in the carriage house. The carriage house became known to hold an oppressive, creepy energy, and was seldom allowed to be investigated. It has since been torn down, due to age and decay.

To accommodate additional guests, a newer building was added. You would think with a newer structure there would be relatively little paranormal activity, right? Well, on "Ghost Hunters" Season 2 again, when Grant was sitting at a table trying to change a battery and the table tilted and jumped – that was in the new building, in Room 1312!

So what does make the Stanley so lush with paranormal activity? There could be many reasons.

First and foremost, it's 100 years old. Secondly, it sits high on a mountain top. If you've read my chapter on the stone tape theory, you know that minerals such as quartz, limestone, iron, and granite have the ability to capture and retain energy, memories, and events from the past. These minerals are frequently found in conjunction with paranormal anomalies and activity. In 2008, the Rocky Mountain Paranormal Group contacted the USGS and had them conduct a survey of the area. Results showed nothing anomalous except high levels of granite in the Estes Park area.

I have investigated locations within mountainous areas and canyons. Many contain deep magma and old lava content. These geomagnetic anomalies can cause high levels of negatively-charged magnetic fields in the ambient areas. High negatively-charged areas seem to coincide with paranormal activity. (Just like Benedict Canyon in Southern California.)

In fact, there are similarities to the Stanley Hotel and Dave Oman's house sitting in Benedict Canyon. Both are built up into the hillside and both the Stanley and the Oman house have rooms with exposed rock and mountainside that act as walls in these rooms. The basement area of the Stanley has huge boulders and dirt halls that served as underground passages for employees during the winter. The Oman house has a service room where half of the room is a dirt mound. Is it a coincidence that these areas are the most active sites on the properties for paranormal activity?

The basement halls in the Stanley have had reports of a small girl's voice and laughter. The dirt room in the Oman house has what I believe to be a strong Native American spirit. I have personally experienced this protective spirit, and the dirt room at Dave Oman's house is one of the most active rooms in the house. Is this proximity to the energies contained in the stone acting as a magnifier for paranormal events? Can having part of a mountain actually inside the building have an affect on the magnetic fields in these buildings?

Maybe it's just the isolation of the place. There is definitely something that has created such a concentration of paranormal activity within the Stanley Hotel. True, there have been reported deaths, as is the case with nearly any hotel. The ghosts of the Stanley are different though. They seem somehow more intelligent. They certainly seem wise to all the paranormal investigators running around the halls, with their infrared cameras and digital recorders. And similar to seasoned actors, they like stepping on stage and giving a show.

Still, it's not known for sure what causes so much activity. I can only tell you my experiences within the Stanley. There are some haunted locations that seem to stay with us, beckon us. They call us back like sirens. The more often we visit, the more welcome we feel. We are drawn back to a location as if we belong there.

Like Jack in "The Shining" finding himself in the old photograph, you truly start to feel as if you were meant to be at these places for a reason, or like you've been there before. The spirits of the Stanley have been there a while.

Don't be fooled though. They know you're looking for them. I think they enjoy screwing with us, and they are good at it! Take Lord Dunraven's room for example. We set up our camcorders poised toward the infamous closet door. It's known to open and shut, lock on its own, and produce voices coming from within. Lord Dunraven in life thought himself quite the ladies' man. He had an arrogance about him in life. This aspect of his personality seemed to stay with him after death. Poltergeist activity is one of the major claims in Lord Dunraven's room. Not only do objects move and break, but if the closet door isn't opening or closing, sounds and voices can be heard coming from inside the closet.

After about an hour with no activity, I started taunting a bit. “We girls are not impressed,” I said. “We know you can hear us.” I got no response. Deciding to move on, we shut off our camcorders. But as soon as he heard the click of it shutting off, the closet door dutifully swung shut and latched! He waited until I shut off the camcorder.

I joined EVP specialists Mark and Debby Constantino in Room 217. While we were sitting in the bedroom, something seemed to be in the bathroom. We started hearing sounds, then voices. Then we heard the sound of running water, as if the faucet had turned on. The sounds were so loud and distinguishable that we thought someone had maybe gone into the bathroom without us noticing. Finally Mark Constantino went into the bathroom to check. Nobody was in there, but the water was in fact running.

We captured several EVPs in Room 217 from what sounded like a male spirit. We asked if he was indeed in the bathroom. We got a “yes” response. We asked him if there was anything we could bring him from the material world. I like asking this question because it always seems to engage the spirits. It’s sort of like asking what trigger object they would like us to bring. We got an EVP asking us to bring him some cigarettes. Guess they can’t hurt him now, right?

I was chomping at the bit to get into Room 1312. I have to admit, I was excited to be in such an active and famous location. I planted myself at that table and waited for furniture to fly! The room definitely has an uncomfortable feeling. It felt like the room had an electrical charge. It’s as if something bad is coming for you and you can’t help yourself but to wait for that knock at the door.

At exactly midnight, the digital clock on a table flickered and a fern rustled from an unearthly breeze.We knew something was gathering energy. My sister, pinned against the wall because she didn’t like the room’s energy, suddenly doubled over in pain.

“I have to get out of here!” she cried. “Something is trying to get at me!”

Most unfortunate for her, she was with me in a room full of several paranormal investigators, which meant we didn't so much help her in her distress. Quite the opposite. In unison, we all turned our equipment toward her and documented what was going on around her with our camcorders.

What? Jeez, she was fine! (She's family, so I can say that)

We sat her on the bed because she was feeling dizzy and we wanted to keep an eye on her. After a couple minutes, her nausea subsided and we let her wait in the hallway. I decided to take her place against the wall. If something had indeed attacked her, I was the closest thing to my sister spirits could get. Besides, I didn't want to subject anyone else to a physical attack.

I began to see the familiar twinkly fairy lights about two feet off the ground. I have seen these types of lights before when ghosts try to manifest. I didn't feel as though I was punched in the gut as my sister felt, but I could definitely feel the energies swirling around me.

This investigation was many years ago, when I was open to interacting with almost any spiritual energy. Today I would never allow my sister or anyone to become a victim of an energetic attack for the sake of evidence. It's not worth it. The energetic forces seemed more interested in manipulating the vulnerable rather than do silly things like tip tables or chairs. So the chair didn't budge for me. The table remained still.

Elsewhere in the hotel, in the room where we stayed, there was something busy puttering around. There was definitely a ghostly visit in our bathroom during the night, turning on water, flushing, and making a lot of noise in the late hours. At first I just thought it was my sister using the bathroom during the night. In the morning she asked me the same thing. It wasn't either of us.

The Stanley Hotel is historic. It is unbelievably beautiful with amazing views and herds of elk on the grounds. It is crazy alive with full-body apparitions, anomalies, and poltergeist activity. If you have the chance to visit, please spend a few days. They even hired a guy who looks like Jack Nicholson to stand around and look like Jack Nicholson!

Don't be put off by the almost hourly ghost tours in the corridors. The ghosts don't seem to mind. In fact, they seem to thrive on it!

CHAPTER 21: Psychic's View of the Stanley Hotel

The Stanley Hotel really is such a beautiful destination that I believe everyone should visit and take the time to enjoy this grand historic hotel at least once. Take a few days and you'll marvel at the incredible scenery around her. I mean, what other extremely haunted hotel has big horn sheep and elk wandering around? It's definitely worth a trip, whether or not you're there to investigate.

The ghosts at the Stanley are unlike ones I've encountered at other locations in America. They can pick up on a psychic in a heartbeat. I have not found anything too negative at the Stanley as far as spirits, but they definitely know that you can see them and they see you.

The fourth floor is reportedly the most active. This is where spirits of children are heard and sometimes seen running up and down the hallway. Sit on the little couch in the hallway during the night and feel the difference in the energy. Regardless of what floor you stay on, you WILL experience some activity.

While some locations have activity day or night, it has been my experience that most of the activity at the Stanley occurs at night. The Stanley definitely has a weird vibe throughout the day, but when night falls, the spirits seem to manifest within the shadows and through poltergeist activity. As a psychic you might feel a bit overwhelmed by all the activity, coming and going. They all seem to come and go so quickly, as if playing a game. As soon as you feel something, have your team start documenting before they slip out and are gone.

There are two areas on the grounds that respond a bit differently. This is where patience, and your openness to how you investigate, will yield better results.

The original building was built in 1909. It features four floors and a brass-walled elevator (which gave me such a bad vibe that I took the stairs every time!). This building has Room 217 – the room many guests, including actor Jim Carrey, fled in the middle of the night. The spirits in this building seem to be more of intelligent hauntings. While the spirit in Room 217 was talkative to us and didn't frighten me at all, the spirit of Lord Dunraven was quite a handful.

Lord Dunraven was a bit of a chauvinist in life and death hasn't much mellowed his prickly personality. Bangs and voices from the closet, swinging of doors, and breaking glass are all in his repertoire. Communicating with him on a psychic level was more about feeling his attitude and recording poltergeist activity. This type of spirit doesn't want to change, doesn't want your help, and considers you to be meddling in his affairs. Document his antics and have fun with this type. They actually are big hams for attention, so feel free to take advantage of the opportunity.

Then there's Room 1402. This room is in a building that is newer than the other structures on the grounds, but it somehow seems to generate the more aggressive energies. This is the room where my sister was attacked by spirit energy, causing her pain and nausea.

So let's go through this encounter step by step and see how you can use your intuition and psychic abilities to enhance the investigation.

When we first entered the room, I could feel an energy. It didn't feel aggressive at first, but we were very unwelcome in that room. I was aware of the energy but was so excited to be in this room – the room where Grant sat right in that chair and that very table tipped. I bounded over to that table and chair like a kid hopping on a roller coaster. Looking back, I should have let my senses feel everything out first, but I guess I was a little starstruck. My next mistake was that I completely ignored the fact that my sister wouldn't even come into the center of the room. She stood pinned against the wall about 3 feet from the door. She looked like she was ready to bolt at any moment.

So the first lesson that I had to re-learn yet again: *listen* to your senses! Don't be so gung-ho to get into a notorious location that you end up tuning out what you know are red flags.

We weren't getting too much interaction with the various K2s, Mel Meters, and other equipment spread throughout the room. What I did notice was that on the nightstand the digital clock kept flashing, changing time, and flickering. The potted fern next to that same nightstand rustled as if blown by an unseen breeze. I looked over at my sister and noticed tiny flickering lights dancing close to her.

This room is reportedly haunted by a ghost cat and the dancing lights were about the size of a cat. I pointed the lights out to my sister. While a ghost cat may not communicate, obviously the energy was strong in the room. This is the time you focus in on who is there. Ask questions. Tell them you are there to learn and want to communicate. Give it a little time. The Stanley spirits are clever and like being in charge. Don't be too overpowering or they might just say screw you and leave – as seemed to be the case in this room. Since we had basically barged into their space, we were not welcome.

It made its point by going after the most vulnerable one in the room, my sister, who isn't an investigator. She began to complain of nausea and pressure. She doubled over and said "I have to get out of here!" Spirit was definitely talking – we just didn't realize right away that it was using her as a form of communication.

We let my sister leave the room and since I had the closest DNA I took her place by the door. I wanted to feel what was there. I have a more assertive energy and certainly was not feeling vulnerable (I rarely do). Spirit felt the difference. More and more flickering lights danced around in the area but I didn't feel oppressed at all.

Whether you're psychic, sensitive, or a paranormal investigator who's completely oblivious and staring at the room through a 2-inch LED screen, what do you do when energy presents itself like this? The best way is to be aware of the type of energy when you enter the room. This energy was not evil or demonic, but it was definitely annoyed. However, annoyed can and did escalate. If you find yourself in a situation like this, let them know you are aware how they feel and mean no disrespect. Remember that you are in their space, after all, and acknowledge that. Establish yourself as strong but respectful. Who did the spirits go after when they became more annoyed? That's right, the most vulnerable.

Don't ever go into a location and let yourself become more and more spooked. They eat that right up with a spoon. Even playful ghosts will have their fun if given the chance. There are plenty of people who are bullies and punks in this world, and you have to remember that they seldom change when they're in spirit form. Come in being respectful. If they continue with the bullying behavior, stick your chest out and just stand your ground. Don't provoke, or you'll just be asking for it. Just stand strong and own the power. You're the one who's still alive.

If they don't want to have any part of you, you should still thank them for the opportunity and tell them you'll be back later. See how the energy feels a couple hours or even a day later. Maybe you caught them at a bad time. Always be respectful, but don't let them run you out. That will come back to haunt you – pun intended.

CHAPTER 22:
Whaley House

The Whaley House in San Diego, Ca

Built on the site of the old gallows. What could possibly go wrong?

The beautiful Whaley House Museum is renowned to be one of the most haunted locations in America. But is it really?

This Greek Revival mansion was built in 1857 by Thomas Whaley. What seems odd now is this spooky haunted mansion now sits smack dab in the middle of touristy Old Town San Diego. You can eat and drink at a delicious Mexican restaurant, walk a block into a 19th century graveyard, then get a T-shirt at a small shop. This is not exactly a spine-tingling atmosphere.

Still, there's a valid reason as to why this location may be so active.

It seems Mr. Whaley bought the prime real estate for a bargain price. How? Well, the land used to be the site of the local gallows. Many a man swung where Whaley House sits. This didn't seem to unnerve Thomas Whaley.

"My new house, when completed, will be the handsomest, most comfortable, and the the most convenient place in town or within 150 miles of here," Whaley said.

The Whaley House also encompasses the courthouse, a general store, and the area's first theater. It was quite a mansion in its day. Alas, even a wealthy family like the Whaleys can have their share of family tragedy. At just 18 months old, Thomas Whaley Jr. succumbed to Scarlet Fever in 1858. The daughter, Violet, divorced and heartbroken, committed suicide in 1885.

But 150 years of courthouse drama and the hangman's noose surely added to the inhabitants. The most notable of those who met their fate at the gallows was Santiago Robinson, also known as Yankee Jim. Convicted of robbery in 1852, he was fitted with a noose and stood up on the edge of a buckboard wagon. As the wagon pulled forward, Jim dangled, his lanky legs almost able to reach the ground. As he struggled, his boots scraping the dirt, he slowly strangled to death.

Are we squirming yet?

So what type of activity were we able to get at our last visit to the Whaley House? We went with a small group of about eight investigators, led by psychic Kimberly Rinaldi.

First we started in the small general store area. It seems Yankee Jim is quite adept at communicating with Dowsing rods. Jim's spirit seems to reside in the general store area and has been known to respond to trigger objects like whiskey or beer. We were able to get several responses from Yankee Jim using the rods. He would move in whichever direction we asked. He seems relatively content at the Whaley House, but we felt since he had died a horrific death that we would bring in an angel for him. We didn't leave the area until the angel placed a hand on his shoulder.

We headed upstairs to the stage theater. We saw some RemPod activity, as it set off alarms and blinked in response to our questions. We began seeing shadows behind the theater's pump organ. Darting and disappearing, several of us saw them. We snapped several photos and were stunned when we saw the shots. In a series of photos, a huge light anomaly can be seen rising behind the organ to a height of about 8 feet. The light was so bright and large that it looked as though the pump organ had caught on fire.

This brings up a valid point as to why we investigate with some light in the rooms. Some investigators insist they have to shoot in darkness because their infrared cameras see in light spectrums that we cannot see. This is true and I agree – to an extent. If we hadn't seen the shadows, we probably would not have taken photos in that spot at that exact moment and would not have captured such amazing evidence. (Plus, I'm a klutz and trip over stuff in the dark.)

The kitchen area is known for having children spirits. I am usually a bit skeptical about child spirits. I have to think that if it is really an innocent child, an angel or loved one would come and carry them home. Sometimes when a spirit tries to convince you it's a child it may be something darker trying to appear more approachable. You should always be cautious with what you perceive as child spirits.

We started an EVP session. "Do you want to come out and play?" I asked.

A long irritated "Noooo" was the only response. I decided I didn't really want to play with whatever that was either; if it was a peaceful spirit, it just wanted to be left alone. Respect a spirit's wishes and move on. You can always return later to see if the mood has changed.

The activity really picked up when we headed back to the largest room within the house – the courtroom. We started picking up a female spirit. She didn't seem attached to the Whaley family, but had felt our intentions to help and stepped in. During an EVP session, we asked if there was anything we could bring her from the material world that she might miss. We captured a Class A EVP several minutes after we asked the question: we heard the word "paper." Perhaps so she could communicate with something familiar?

On this investigation, our group was made up of mostly empaths and psychic mediums, so we decided to put away most of our equipment at this point and try to communicate psychically. We felt another young girl approach us, and could tell she desperately needed help. This spirit was almost in a panic state, scared and crying. She had died of an illness but seemed to feel the burial process of them wrapping and binding her in burial bandages. Imagine how that must have felt, trapped in that mindset and terrified.

Often when spirit tries to communicate through me, I get these awful coughing attacks. I had been fine all night and now suddenly I had tears running down my face and I couldn't utter a word if I tried. I did the next best thing and continued rolling on my digital recorder.

While I was trying not to succumb to a coughing attack, psychic Kimberly Rinaldi started verbally communicating, telling this scared young girl that what she felt was only a memory. We did our best to calm her down and ease her pain.

To comfort spirits in this much need, we call in angels to help us. I mention this because when we do this, blue sparks can be seen flashing from our fingertips. A witness at the Whaley House, who was not familiar with this, saw the sparks and asked what they were. These blue sparks have been photographed on several occasions. We brought in angels to help guide the girl where she needed to be.

As visitors are so open and inviting of spirits at noted haunted locations, these locations become somewhat of a portal. I think spirits who find comfort in a location stay, while others are attracted to the spiritual energy and come in seeking aid or sometimes simply acknowledgement. Nobody wants to be forgotten and I think sometimes a simple conversation is all they really want.

So the next time you're in Old Town San Diego in California, stop by the Whaley House Museum. And while you're walking down the street, look down at the brass markers on the sidewalk. They say "grave marker." When they expanded the streets, they paved over part of the old cemetery, placing this brass makers wherever a body is still buried.

Like many historical places, Old Town San Diego has a lot of secrets.

This light anomaly was captured at Whaley House. This glowing light anomaly grew over the course of several frames. There was no light source behind organ, no reflection.

CHAPTER 23:
Psychic's View of the Whaley House

The Whaley House Museum has been an icon for the paranormal for years. It is one of the best known locations on the West Coast. The fact that it sits on a very busy street in the tourist mecca of Old Town San Diego makes it a big draw for not only the serious investigator but the curious snapshot-taking tourist as well.

I commend the Whaley House for capitalizing on its haunted history. These experiences are needed for the maintenance of the museum. The noise and flow of visitors can make investigating the Whaley more of a challenge, but most investigators know how to accommodate that sort of environment.

Because tours are held regularly at the Whaley House, some of the regular spirits seem more like trained monkeys in their interaction. True, spirits are the money machines in a haunted location, but to ask them the same questions 20 times a day is doing a disservice to everyone. We need to use every opportunity we have with spirits as a chance to comfort, heal, communicate, and bring peace to the deceased. We aren't doing what we need to propel the field forward if we trap a popular ghost and bring them no closure. It's cruel.

I don't mean to snub what these groups do. But I do think we need a more flexible approach to locations who have "resident" ghosts.

Take Yankee Jim for example. He's tired and feels he doesn't deserve to cross. He has created his own fate because of his actions on Earth. When we were investigating at the Whaley House, we brought in an angel for him anyway. We gave him a choice. If Yankee Jim decides to move on, I believe he can return later when he resolves his own issues. Many people, when they die – especially when they are executed – continue to have emotional issues that remain unresolved in the afterlife. Wouldn't it be better to try to give him a chance at absolution and pcacc?

In the kitchen area, we did a quick EVP session. We asked if the children wanted to play. We received a clear no response. Rather than push the issue, we left them alone. The energy felt OK, but the EVP told us they wanted their privacy. This is when partnering conventional investigative techniques with your psychic skills comes into play. You won't always get a bad feeling when they want you to leave. It happens.

We visually saw the shadows behind the pump organ, but we felt something there too, which made us look there in the first place. If you feel compelled to look in a particular direction, immediately have your team take photos. There's a reason you are looking in that direction.

Our psychic abilities really began to shine when we went back to the old court room. We picked up on spirit entities sitting in the corner. They were just observing. We started an EVP session. We picked up on the poor scared girl. She seemed to know she was dead but was experiencing the feelings of having her body prepared for burial. Those feelings seem to terrify her. My throat started tightening up. This happens when a spirit is trying to communicate. It starts affecting my vocal cords. If this happens to you, it's because a spirit is using you to try to communicate by manipulating the vocal cords.

We kept telling her that what she felt was only a memory. We asked her if there was anything she missed from the physical world. We got a response from the recorder. Not only was she manipulating my vocal cords but was she also asking for paper so that she could write what she wanted to say. This was a strong and experienced spirit.

Talking to a spirit you know is in extreme distress has several advantages. They are eager to communicate so you're likely to get some quality EVPs. But while you're asking them questions, it keeps them busy while you're beginning the healing process. Bring in angels or any higher vibrational energy with which you are comfortable working. Ask whatever familiar spirits who might be around to please reach down and grasp the spirit's hands; you don't need to know their relatives in order to call on them for help in this way. Just make it sincere, and it will work.

Haunted places will usually remain haunted. Such a place will not become devoid of paranormal activity by doing healing work. As one moves, the light and energy of a paranormally-active location will attract other spirits. We won't run out of dead people. There's a difference between evicting a troublesome spirit that traumatizes the living, and actually moving spirits from a paranormal location. These spots remain so because they feel welcome and will usher in other spirits like revolving doors. A spirit like the little girl at Whaley House wasn't associated with the land; she came for help because she was attracted to the light we brought the location.

The 30 foot metal staircase in the Oman House. Geomagnetic anomalies, Native American battles, tragic deaths of Old Hollywood. Could this be an antenna to the Other Side?

CHAPTER 24:
Oman House

In order to understand the uniqueness of the Oman residence on Cielo Drive from a paranormal standpoint, we must attempt to understand the location and its rich history. Located up in the Hollywood Hills – above the Hollywood sign, above the famed Hollywood Hotel – Benedict Canyon is a place plagued by an abundance of hauntings.

Benedict Canyon was home to Hollywood's elite, whose lives soared, only to come crashing down all too prematurely and often tragically. Just a stone's throw from the Oman house sits what remains of Falcon's Lair, home to Rudolph Valentino. Valentino was the Silent Era's glamour boy, dying much too soon. Falcon's Lair is reportedly extremely haunted. Owner Pia Zadora had the home torn down to its supports because she was so traumatized by all the paranormal activity inside. Now just a shell of the original home exists.

This canyon is where television's first Superman, George Reeves, was found dead from a gunshot wound on June 16, 1959. Initially dubbed a suicide, it seems something sinister was involved and a cover-up by authorities was afoot. (Pick up the book "Speeding Bullet: The Life and Bizarre Death of George Reeves," the second edition published in 2007, for more details about that.) Reeves' fingerprints were not found on the gun, nor was any gunshot residue found. Additionally, several bullet holes were found in the floor of his bedroom, hardly depicting a suicide. His case was never officially solved.

The history of Benedict Canyon turns ever darker on the night of August 9, 1969. A home at the address of 10050 Cielo Drive became the story of nightmares. Seeking revenge on music promoter Terry Melcher, Charles Manson sent members of his "family" to wreak havoc and bloody mayhem. Instead of Melcher, the home was rented to Roman Polanski and his wife Sharon Tate. That night, Steven Parent, Abigail Folger, Wojciech Frykowski, Jay Sebring, and a pregnant Sharon Tate were brutally stabbed and murdered.

Chants of "Someone's going to die tonight" still echoes in the wind in Benedict Canyon sometimes.

But why all this tragedy? Why this canyon?

It's no surprise that the U. S. Geological Survey (USGS) lists this canyon as a geomagnetic anomaly. Although the USGS states that the hills are geologically too young to have either large deposits of iron or magnetite, Benedict Canyon still emits strong, positively-polarized magnetic fields. This not only makes for a fun-house effect, but can render instruments useless. Large magma deposits were also discovered during construction, adding to the magnetic mayhem.

Construction of the Oman residence was quite a challenge, to say the least. Built into the canyon side, it required huge support beams driven into the ground. It is four stories, stepping up the canyon. Remember all that magma and magnetic polarities – and then they drive metal support beams into the ground. Add a 30-foot metal spiral staircase and that's what I call a paranormal antenna!! Many psychics and sensitives, including myself, feel quite the fun-house effect going into the Oman residence.

Dave Oman's house first gained notoriety when featured on "Ghost Hunters" in 2007. It was here that noted psychic Chris Fleming introduced Jason Hawes and Grant Wilson to the home and showed us how to use a K2 meter for simple yes or no communication – a poor man's electric Ouija board, if you will. In the episode, we all remember seeing Tango get the chills when he receives his first response from the K2. This footage also showed us Chris Fleming validating through spirit communication that Jay Sebring was indeed one of the spirits inhabiting the home. Localized temperature drops on command and EVPs were also documented.

Six years later, I had the privilege of joining Chris Fleming on a return visit to Dave's home in February 2013. He invited only six of us, and even though I was running a high fever I was not going to miss this. Dave greeted us in his bouncy, gracious manner, and gave us a tour of the hot spots. A cameraman documented most of the evening's events. (Many clips can be found on Youtube – search "Christopher Fleming Oman house.")

The entire Benedict Canyon also has strong Native American energies, and in fact was the site of the last Native American battles with the American Cavalry.

Dave's house has a utility room that is exposed to the hillside, with a dirt mound in the room. This room seems to be the center-point of that extremely strong Native American energy. Dave said several psychics have picked up on the energy and had also picked up on his horse spirit.

Chris and I started an EVP session. Chris asked, "Can we hear the sound of the horse, please?" When we played the recording back, both of our recorders had the familiar clip-clop of horse hooves. What's really unusual is that this room has a dirt floor. Our feet could not have made that noise, and we stood very still.

What's even more crazy is we received the EVP of the horse hooves *before* Chris even asked the question. This gives credence to the theory that spirit can communicate telepathically. They knew what we were going to ask before we even asked it.

One of the most rewarding moments was during a Spirit Box session. Dave Oman handed me a handwritten letter by Sharon Tate to her mother. As I read it aloud, I couldn't help but feel a strong connection to her. We asked if she could talk to us. I had this uneasy feeling that something was holding her back. The voice that came forth was a distraught female voice.

"Can't..They won't let me."

To explain what happened during one of the EVP sessions, you have to understand some of Chris's methodology.

Chris's methodology is very similar to my own, but he works more closely with angels while I work with more Native American energies. His psychic and mediumship work has him partner strongly with angels. Now, I am not here to provoke a religious debate. I have personal experience with assistance from the other side of the veil. What many call angels I think of as *very* high-vibrational energies.

Chris had been doing a short EVP session and in the playback you can hear spirit say, "Oh my God." In the next EVP session, you could hear, "I see angels!" Chris teared up knowing that the angels who guide him could be seen by spirits. The experience was amazing.

Later in the evening, just three of us were in a small room conducting more EVP sessions. On one of the responses, we heard someone ask for help. It is a moral obligation to try to help those spirits who ask for help, but it's also important to offer help when they are unable to ask. This is when we say a prayer for the dead. It can be any sincere, heartfelt request for those above to please come and carry them home.

Chris had his recorder running when we gave our prayer. We requested that the angels please reach out and take the hands of their loved ones and help to carry them home. Of course, we were crying. (And if you're doing it right, you'll probably cry too.)

When Chris played back the prayer, we could hear thank you's scattered throughout the recording. (More tears, of course!) It's that kind of relationship with spirit that we all should strive to achieve. The purpose here is to communicate, get answers, and, most importantly, help those on both sides of the veil.

To sum it all up, on our first investigation at the Oman House with psychic Chris Fleming, we were able to capture what we think is the voice of Sharon Tate while using a Spirit Box. We had EVP sessions that told us they were able to see the angels that protect and guide Chris Fleming. EVPs of voices saying "Thank you" were recorded while saying a prayer to help carry the trapped spirits home.

With all of this evidence, of course I was delighted when Dave Oman invited me and my team back to his home for a subsequent investigation. What would our next visit to Dave Oman's house bring us?

The Oman residence is nothing if not consistent in its paranormal activity. Furthermore, Dave Oman has a unique relationship with the spirits who come and go throughout his property. He has a genuine respect for the spirits and not only communicates with them in a present tense as if they are family, but refuses to let visitors or investigations refer to those murdered as merely victims. He understands them in a way that makes them feel like they are in a safe space.

I think this is important to remember while doing any investigation. Spirits are not there for our amusement. Spirit is not there to be provoked and taunted so that we'll have some awesome evidence to post on Youtube. These were people's family members, so you must treat them with respect.

That being said, there are multiple levels of energies within the residence and throughout Benedict Canyon. Our group is comprised mostly of psychics and sensitives, so we can see what's out there. The canyon holds dark elemental energies that flow like slimy leeches in some gruesome river at the bottom of Benedict Canyon. As mentioned, there is also a strong, protective Native Spirit energy that watches over the darker energies and keeps them at bay.

As Dave walked us through his home, looking out onto the canyon through huge, expansive windows, we could see the protective spirits watching us. They looked to us like scouts hiding behind the rocks, peering out. If you know that an earth energy is present, bring an offering. It's a goodwill gesture. We brought some sweet tobacco and sage. As we placed the offerings, several of us felt the knowing pins and needles tingling on our sides. We felt that this was spirit getting to know our intentions and our motives. I found this response similar to a welcome, which was a good start!

We did a short EVP session. What we captured was shockingly similar to what we had captured in this location during the previous investigation. Once again we captured the clip-clop of horse hooves, and once again the EVP occurred *before* we asked the question. We have since captured this same EVP a few times now.

The most active part of the house for poltergeist activity is on the top floor. There sits a huge aquarium festooned with figurines from horror movies. Of all the figures, a battered Beetlejuice figure takes the brunt of abuse; it gets knocked over on a daily basis. We set up a static camcorder to document, and set a RemPod close to Beetlejuice. As Dave is recounting past experiences, the RemPod begins alarming. What's funny is that the RemPod seemed to alarm in response to Dave's remarks. We unplugged the aquarium lights and left only one filter running, which allowed us to debunk any disturbances of electromagnetic fields.

The entire time we were investigating, the RemPod was merrily beeping and alarming all the while. Walking through the house, I was drawn toward a long, dark hallway, so I followed. I felt compelled to place my hand against the back block wall, which is a wall that butts up against the hillside. The sensation was so weird – it felt like something was moving on the other side, trying to find a way in.

We conducted a quick EVP session at this area. When I asked if they were trying to get in, psychic Kimberly Rinaldi and I both heard laughing. And it wasn't exactly a pleasant laugh either. It was a creepy, evil clown-type laugh.

We got nothing on the recording. We played the recording back and, as we were discussing this evil laugh that we had heard with our ears, the lights in the hall started flickering on and off. Dave checked the switch and assured us that the lights didn't normally act like this. Indeed, they all lit up on one panel, not able to intermittently flash on and off. As soon as we stopped talking, the lights all went on, and finally stayed on. During this entire time, our RemPod was regularly beeping upstairs.

We ascend the staircase to find, sure enough, Beetlejuice is toppled over. Rewinding the tape we see something interesting. Just before the Beetlejuice falls, the fish in the tank start swimming frantically, as if something is after them. The RemPod surprisingly did not alarm when Beetlejuice fell. Why spirits abuse this one figure remains a mystery, but it has been recorded countless times falling while nothing else is disturbed atop the aquarium.

I feel bringing sensitives and psychics this trip helped us connect with the protective energies more than specific human spirits this visit. The Oman residence contains multiple layers of energies and many spirits. It also has unexplained magnetic anomalies in areas of the house. This house is consistently active, thanks in part to Dave Oman's respect and acceptance of these spirits. This makes the Oman residence a valuable training tool for multiple investigations. By conducting multiple investigations over a period of time, we hope to find what makes paranormal activity happen and when. We can study the conditions under which paranormal activity occurred while controlling as many variables as possible.

I humbly thank Dave Oman for his hospitality and hope to return soon.

CHAPTER 25:
Psychic's View of the Oman House

Dave Oman's house could almost be considered a training ground for paranormal investigations. Factors such as the openness of the homeowner, the ties with Native American energy, and the geographical anomalies – to say nothing of the horrific crimes that took place there – all add to the nurturing of the paranormal.

Most investigators completely canvas the interior of Dave's house with myriad infrared and night-vision video equipment. The home yields poltergeist activity as well as light anomalies. But let's reexamine the investigation with Chris Fleming, and review how a psychic's input enhances this type of location.

Because of Benedict Canyon's strong Native American ties, it behooves someone entering such an area to engage the energies and spirits with a high level of respect. This applies in any area deemed sacred, be it a rock formations, burial grounds, or even an area where certain sacred plants grow. Don't disrespect what they deem sacred – even if it's not a belief you subscribe to.

It never hurts to bring some sort of offering. Even if you aren't sure if it's considered a sacred area, it shows a positive intent and respect – and that's always a good introduction.

Chris and I started in the dirt room. We wanted to introduce ourselves and let them know our intentions by showing respect to the Native American energies first. These types of energies are traditionally very established and deserve top-priority acknowledgment. It's like starting off on the right foot, so to speak.

The dirt room usually yields little in the way of physical evidence. It's dark and small and, aside from having strange magnetic pulls, it doesn't draw investigators who prefer photographic evidence. It's great for the sensitives though. The protective spirit that inhabits this area will often wrap around you. I believe it's a way of greeting you, like a blanket. I find it very welcoming, but some people react to being touched as intrusive.

Remember in "Ghost Hunters" when Zak was in this room on the ninth season premiere episode? The spirit energy did exactly this and Zak bolted out, assuming it was something negative. Take a deep breath and feel. Does this touch feel harmful? Does it feel like an arm around you, or a does a touch feel more like you've been groped? Know the difference and communicate your impressions to members of your group.

Chris and I asked to hear the sound of the horse during our EVP session. We tried to associate to something that was important to that spirit. Through our introduction we had established a respectful relationship. This was validated by the fact that we received an EVP that answered our question before we asked. This indicates rapport and telepathic communication. We have received this EVP in the dirt room several more times and each time the sound of hooves is heard before we ask the question.

Chris comes to an investigation with an energy and compassion that is contagious. By watching and listening to him interact with spirits, it's as though these spirits are some long-lost friends. I think the spirits pick up on that and are open to communicating; they feel like it's a trusting interaction.

Like Chris, I have found that doing short, or burst, EVP sessions yields the best results. Feel who might be there. Start asking a few questions. Don't go too long – just a couple minutes or so. Then playback the recording and see if you captured any voices. If you get an intelligent answer to a specific question, great. If you get maybe just a word but can identify emotion or gender, go from there.

This is a conversation. You ask and answer, they answer and ask. If you did hours of recording and took it all home to review days later, you would miss out on these conversations and interactions. When we got the EVP of the spirit seeing angels, we would have lost something if we hadn't played it back in the moment.

My favorite part was when I read the letter written in Sharon Tate's hand. Psychometry is getting messages or information from holding objects. I find it very useful in connecting, especially when a location can have a lot of swirling energies. Holding something that was dear to them focuses the energy. When I read that letter, I just started tearing up. I never met Sharon Tate and I never met Sharon Tate's mother, to whom the letter was written, yet I could feel the sentiment. I could feel Sharon with me for that one brief moment. I was picking up her emotions and could tell someone – or something – was holding her back. She wanted to talk but something didn't want her to. Feeling a connection makes EVP sessions or Spirit Box sessions even more successful.

Walking through the house, there are spots that feel protective, areas that feel malevolent, and some places where you just feel dizzy, as if the room is spinning. The hallway always feels like something bad is at the end of it, and yet I feel compelled to always check it out. Many weird light anomalies have been captured in this hall, twirling rod-like lights that manifest from one wall and disappear into another.

At the end of the hall is a cinder block retaining wall that butts up to the hill. It is the only thing between what's crawling around out there and the relative safety of the house. I felt compelled to put my hand on the wall. I had to feel what was on the other side. Remember the slimy, leech-like elemental energy I mentioned that could be found at the bottom of the canyon? I felt them here. It was as though I could feel them "swimming" through the dirt. What was even creepier was that Kimberly Rinaldi and I could hear strange, clown-like giggling.

If you feel something like this, have others place their hands on it. Don't tell them what to feel, just let them experience it and ask them to describe what they felt. Touching areas you feel drawn to gives your sensations validity. If you are drawn to a picture, a particular headstone in a cemetery, a tree, or any other object that you are able to physically get ahold of, place your hand on it and listen to its story.

The 100 year old Graber House in Ontario, Ca

Protective Elemental energy AND a Creeper!

CHAPTER 26:

Graber House.. Angels and Olives

Let's face an unpleasant truth. There is much in the realm of the paranormal and the metaphysical that we just don't understand. There is stuff out there that is not only frightening, but harmful. There are entities, spirits, and energies that can hurt us. There are things not of this world that can manipulate matter and energy in order to inflict physical pain, push, scratch, pinch, and cause abdominal pain and nausea.

We teach ourselves and our children to be wary of people we don't know. We talk to them about "stranger danger." But what about the things we cannot see, the things that go bump in the night? Who protects us from them?

Investigating a haunted location is similar to preparing for battle. We prepare by researching the history of a location's past and interviewing witnesses who have experienced something they could not explain. This information gives us insight as to who or what may be causing the reported activity. Detailed accounts of events can tell us if what they are experiencing is something resembling a residual haunting – which is a ghost or spirit energy that is just a memory, like a piece of film playing over and over with no conscience and no ability to interact.

If witnesses are able to capture EVPs, or spirit voices caught in the white noise of an audio recording, or if objects are moved and disembodied sounds are heard, then more of an intelligent haunting may be present. It may be that of a loved one. Perhaps Aunt Polly is still puttering around the kitchen, making sure we copy her recipes to a T.

If such a spirit is present and attached to the property, their energy may affect us in a wide range of ways, from uneasiness to curiosity, to even welcoming or feeling comforted by their presence. I know my dad pops up every now and then, and I hear his voice and feel his familiar presence. I find it comforting.

But there are spirits and energies that may become protective and territorial. Entering into these locations will give you a definite feeling of uneasiness. They don't want you in their space and will let you know about it. Have you ever just gotten the creeps going into a place for no real reason? Have you been on an investigation and had continuous equipment failure, batteries drain, cameras malfunction, static and interference? You are in their territory now. You never want to be the bully, but you don't want to get pushed around either. This is when you practice claiming your personal space and standing your ground. Don't be arrogant or rude about it; being pushy doesn't please anyone, alive or dead.

Ignoring these fundamental guidelines is what gets paranormal investigators into trouble. That kind of attitude is what gets folks scratched, slapped, and pushed. Dealing with aggressive energies or negative spirits is a lot like dealing with people with similar personalities. Don't get pushed around but don't intentionally antagonize them either.

The Graber Olive House is the oldest surviving business in Ontario, California, family owned and operated since 1896. The property consists of barns, storage buildings and sorting rooms, most dating back to the late 1800s. Enormous storage rooms that hold concrete vats for curing olives look much as they did at the turn of the century. It's a business where most of the work is still done by hand, much as it was done 100 years ago.

The land has remained virtually unchanged for over 100 years, all the while new housing projects rose up around it and land was bulldozed and carved out to accommodate the growing area. This means a lot of changes took place in the energy around it; families came and went, memories were made and left behind.

I believe that the Graber House is becoming some sort of vortex for spiritual energy. It still holds old earth energy from Native Americans, early settlers, and even people who worked for years at Graber, sorting and packing olives for decades. This unchanged plot of land has become a place of familiarity to spirits. To these entities, it's a place they still recognize. I think that's another reason historic locations remain haunted for many years. If they remain relatively unchanged, spirits will seek out what's familiar to them. A ghost from the 1800s may stay as well as a ghost from the 1940s. It's as though the memories become layered. Some locations can contain layers and layers of memories.

This type of energetic stability in these old locations has been known to attract lower vibrational forms too. Some are known as creepers. The Graber has one of these horrid creatures. A creeper is what I refer to as an elemental. It is something that has never walked the earth in a human form. They are not demonic, but they are also not a pleasant thing to encounter. (Think Gremlin.)

Creepers have been seen in many haunted locations. One that is frequently encountered is at the Trans Allegheny Lunatic Asylum. Creepers appear as a dark mass, sometimes changing form. They crawl up walls and across ceilings, and may also pass through walls. Clinging to the ceiling like some alien spider, they are the stuff of nightmares, and I've seen one.

At our first first investigation at Graber, I was already aware of this creeper energy and I admit I was curious to encounter such a thing. Seeing a big, dark mass creeping spider-like up the wall, I wondered if I would be able to communicate with such a thing. Would I want to? Add to that a bunch of 100-year-old buildings and a large, wooded clearing circled by ancient Sequoia trees, and it sounded like an awesome investigation was about to begin. I was with a team of about 15 investigators along with EVP specialists Mark and Debby Constantino. I was looking forward to an interesting night.

Earlier in the evening I had investigated around the vat rooms, taking some still shots. It was October and the vat areas were closed off as they were gearing up for the olive harvest and curing. If this was where the creeper resided, I couldn't sense it. I wasn't getting any of those vibes – the hair standing up on the back of the neck, or that feeling of being watched. It actually felt pretty quiet.

We headed up to the barn loft to start our EVP sessions. Although it was fall, California weather is mild, so it was still quite warm in the loft even at 10 at night. We ran several digital recorders simultaneously and did short burst sessions. By asking questions for only a couple minutes and then reviewing on the spot, not only is it a lot less tedious than listening to hours of audio later, but if you get a spirit voice responding intelligently to your questions, it's immediate validation and helps direct the line of questioning.

We were able to capture several Class A EVPs or spirit voices that seemed to give us intelligent responses to our questions. When we asked for names, we heard "Robert" in the white noise when we played back the recording. This intelligent communication continued for about 20 minutes.

Sitting in the darkened room trying to adjust to the dim light I noticed something moving above us in the dark – something big. I looked up to see what it could possibly be and right there was the creeper! It was easily 10- to 12-feet long, blacker than the darkness around it, with four long, spindly arms stretching another 10 feet past its body. It had a large, human-like head.

But it was its movements that were the most horrific. Its small feet made a suction cup-like sound. *Thwp, thwp, thwp*, across the ceiling. I swear I could hear the *thwp, thwp, thwp* of its suctioned feet! It slowly rolled its huge head backwards and looked right at me. I think it may have grinned. I wasn't sure what to do. I was almost afraid to mention its presence for fear that calling attention to it would cause it to drop down on us like some nightmarish spider. I gulped and kept watching.

It watched me. It rolled its head backwards in an unnatural twist.

Thwp, thwp, thwp, across the ceiling, back and forth. If anyone else saw it, they weren't saying! I know it sounds weird not to tell everyone, "Hey, look up!" but something inside me just told me no, leave it be, and I think it was enjoying its ability to intimidate.

While I still wasn't quite sure what to do, my gut feeling was to ignore it as best as I could. I think it was hoping I would call attention to it. I know I'm there to investigate and document, but as a psychic I just knew not to acknowledge it. I just don't want to engage with such an entity. I didn't need a big alien spider thing dropping down on me – no, thank you. I kept my eye on it the whole time I was in the loft, believe me. I made sure I didn't get directly underneath it. Having such a prolonged encounter with this type of elemental being was a bit unnerving and something I will never forget.

Everyone decided to head down to one of the packing rooms. Pallets of labels and aluminum cans lined the walls. A rolling conveyor belt sat silent. It was a rather utilitarian and impersonal workroom, rather non-descript from a paranormal standpoint. But looks can be deceiving. There was definitely a presence there, an energy that felt heavy and oppressive. Something did not want us in that room. I had a feeling that something was being quite territorial and wanted us out. It would soon make itself known and make sure we got the message.

Mark and Debby came in and sat in the middle of the room while the others gathered around them to start another EVP session. I walked to the far side of the room. I almost immediately started feeling a presence, and an unmistakable pressure against my back like someone was watching me. We weren't provoking and we were just present in the room, so I couldn't figure out why this presence was so uncomfortable with us being there. The room was a workroom, filled with utilitarian objects, tools and supplies – certainly not the sort of thing that should make a spirit territorial. Nonetheless, something didn't want us in there.

I started feeling a different kind of pressure against my back, this time like someone was pushing me. We weren't taunting or provoking, and

I wasn't about to be bullied. There are people who bully and harass others in life, and I think some of those personality traits carry over to the other side. Unrestricted and free of the confines of a mortal human body, these aggressive personalities are free to create havoc without consequence. They still enjoy the power trip.

This is how I deal with a pushy entity though: I push back. I push back in a passive aggressive way, by utilizing some techniques I learned when training horses. I turn my back to whatever I want to move away from me and simply back up. I walked backward and pushed this entity back into the corner of the room. By avoiding eye contact, you can appear less aggressive. I didn't push it all the way back, but I pushed it away from us and out of our space.

It stayed back for a minute or so. Then I felt more pressure against my back. Really? I backed up again. I don't always recommend this type of action, but don't be bullied out of a room. While on some level it may appear insensitive, touching a person is indeed a form of communication for spirits, so you can tell when they're being territorial and bullying. If you allow yourself to get pushed out in this manner, it makes you appear vulnerable, and that's a bad road. I'll tell you what happens when you don't handle this type of energy in the right way.

I looked over at the reporter who had joined us. All the color had drained from her face and she was sweating like crazy. She looked as though she would pass out. She doubled over in pain. Before I could catch her, she just slumped to the floor. Feeling a little shocked and perhaps embarrassed, she said, "I'm OK, I'm OK." But she clearly wasn't OK. Clearly whatever was in the room with us was manifesting physical attacks on others in the room. Now dizzy and nauseated, the reporter was helped to her feet and we got her out of that room. We sat her down outside in a walkway and gave her some water. Being away from whatever was in that room, she began to feel less queasy and returned to normal.

But I had to go back into that room. I needed to find out what was really going on in there. Was her physical illness actually caused by some negative force?

As soon as I walked in, I felt the pressure in my back again. I instinctively backed it up again into the corner, this time a little farther into that corner. I attempted to communicate with whatever was pushing against me in an effort to stop this shoving match. I told the spirit or entity that we were not there to make it leave, but we were going to be in this room and we would leave soon. The pressure subsided.

I looked over at psychic Kimberly Rinaldi and noticed that she, too, was now starting to double over in pain, and looked very uncomfortable. Kimberly was beginning to feel the effect of this entities' bullying behavior. Kimberly's eyes grew wide as she looked at me. We both knew what was causing people to become suddenly ill. Wincing in pain, Kimberly had to get out. Clearly it was making its point. I decided I would give it back its room, but on my terms. No creepy elemental was going to chase me out. I stood my ground until I felt no pressure on my back, and then calmly left the area. Well, sort of calmly. Let's just say I walked out pretty fast.

So the three of us are now sitting in the breezeway, taking a water break, and clearing our heads. Without provocation, this entity manifested physical attacks. We wondered how the 10 or so people left in the room might be feeling. Would another person be attacked? How could they not be feeling it?

Suddenly, another investigator came rushing out, his face just ashen white. He was clearly scared. He was bringing his mother out, and he was having to almost carry her, she was that distressed. She was barely able to walk, twisting in pain, and she seemed to be having some sort of seizure. Both were clearly shaken.

Was this seizure caused by this entity? Why had it attacked her? Was she an easy victim, vulnerable to the aggressive energy that resided in this packing room? Thank goodness that within a few minutes, she was feeling more herself and able to move more easily. Like some ghostly junkyard dog, it was going to make sure we left its room and left it alone.

We were definitely not as prepared as we should have been for Graber House. Our next visit would require better boundaries and more protection from above – I'm talking angels, and lots of them.

Months had passed since we had last visited Graber House. Our next visit was in the spring, when the olives are still growing and the sorting and packing equipment is silent. The vat room was just rows and rows of empty concrete curing vats.

This time was going to be different. At this investigation, we would be bringing guests investigators to Graber House. Some of the folks were experienced paranormal investigators, while some had never been on an investigation before and were curious beginners. Most knew a little about the spirits that resided at Graber House, and like most other paranormal investigators, they were out to find the creeper.

Going into a haunted location with the intent of provoking an aggressive spirit is simply asking for trouble. No matter how protected you think you are, it's just never a good idea. If a spirit or lower-vibrational energy takes a dislike to you, they *can* hurt you. I've personally witnessed people become suddenly violently ill, throwing up, and developing stomach and chest pains. I've seen people get pinched, poked and had objects knocked over in an effort to trip them in the dark. We decided if this creeper or other aggressive energy presented themselves, we were not going to engage any type of interaction with them. It could just be too dangerous and I just don't like letting them play games on their terms.

While some of you might be thinking that it would be cool to be poked or pushed by some unseen entity, the true danger lies in the fact that sometimes these entities will attach to you and, yes, follow you home! No, this is not fun. A clingy negative energy is like a sickness. It can affect your mood, your health, and your relationships, and never in a positive way. Sometimes it can be so profound as to develop into mental illness and depression. Don't risk it and know how to extricate one if you have to.

We made a decision from a safety standpoint. Both Kimberly and I decided that for this investigation we were going to bring in higher vibrations. Doing this can be as simple as reciting a prayer beforehand to ask for protection, but we wanted to take it a step further. We stood together and called in our angel guides. We wanted the big guns to protect us on this trip. We had not only our own safety to think about, but the safety of our guests, some of them beginners. Their safety was paramount, and we wanted to ensure a good first experience for the beginner ghost hunters.

Feeling confident in our protection and ready to have a good investigation, Kimberly and I drove up toward the location, but first needed to pick up Kimberly's sister along the way. As we were rolling along the freeway, we both started hearing the distinctive *clink, ching, clink*, of loose change falling onto a hard surface. *Clink. Plink. Clank.* What? The car floor is carpeted. *Clink.*

"Um, Kimberly?" I said. "Are we manifesting coins out of thin air again?"

Clink.

"OK, pennies from heaven," Kimberly said. That's a good sign that someone is watching out for us from above.

We pulled into her sister's driveway. As I get out of the car I look at the driver's side floor. There must have been six or seven quarters and two pennies. (Now that would be perfectly normal for my car, but her car is always immaculate.) Pennies she had encountered before, but the quarters especially puzzled her. Pennies would manifest whenever her dad was watching over her. Kimberly had received many pennies sent from her dad's spirit. But quarters?

Kimberly's sister snickered. "You asked Dad for pennies? I asked Dad to send me quarters! He always liked me better!"

Graber House consists of several buildings and storage areas connected by walkways. They were close enough that if some investigators split up and were on one part of the property, it was most likely they would unknowingly contaminate any audio other investigators were recording in another area. So we decided we would do all of our EVP sessions as a group to lessen the likelihood of audio contamination. Later, when people start getting the itch to investigate on their own, they could roam the grounds to take photos and get their own feel for the place.

Being secluded in a haunted location, you really get a feel of where it seems creepy, and what corners seem foreboding. When you're alone in that dark shadowy room and you see a darker shadow growing in the corner, the hair standing on end and every tiny noise amplified, that's when you better have your cameras ready! That's why it's important to have someone with you – for safety, as a witness, and to laugh at you when you pee your trousers.

So before we even started the evening we gathered everyone around to get all of us in the right frame of mind. A good exercise for this is what is called toning. Toning is a lot like chanting. This vocalization is especially good when you can lower your voice and really make your chest vibrate. By having all of us do these deep, chanting tones, it creates an equality of vibration within all of the participants. Sound is the closest vibration to physical matter; toning breaks up stress and clears the mind. Standing in a circle and holding hands, we did three low chants.

We've experimented with placing a digital recorder in the center and recorded while toning. Try it, you'll be amazed at the EVPs you'll get. Spirits seem to be very curious about all this noise and we've been able to record some cool EVPs while toning. The spirit voices sounded like very excited chatter. Perhaps they're simply wondering about what it is we're doing.

I don't know if it was the toning or the fact that we invited higher vibrations to join us, but it was like many little eyes were watching us from the trees and the bushes. I can't say it felt threatening, more like a very earthly, ancient woodland creature watching us curiously. It's hard to describe, but it was as though the land itself had noticed our presence and was watching us, and indeed curious about what we were doing. Several of us noticed darting shadows and firefly-type lights flickering about.

While the feeling of the outside spirits seemed curious and friendly, that energy changed on a dime when we entered the vat rooms. There was a very unwelcoming vibe, like having some territorial guard dog staring you down. Kimberly and her sister and I had noticed one back corner that was especially creepy and oppressive. While the others took a few minutes to snap some photos, the three of us went back into the corner. We wanted to know what kind of spirit or entity we were dealing with.

All of a sudden, we noticed the air just seemed electric, like a static charge. This one corner had a tense, angry feel to it, and then we noticed the room's peculiar smell, like that of sulfur and dog feces. That sudden smell of bile can be indicative of a demonic presence. We don't play that game. I have no interest in engaging with demonic energies. Why give them the satisfaction? What they want is to be the center of attention. It empowers them. We don't need that kind of drama – which is why we have personal and spiritual boundaries.

This vat room area is where the creeper is usually seen and photographed. While I don't believe this creeper is something demonic, something that crawls up walls and looks like some twisted alien spider is certainly not a long-lost loved one either. This whole vat room area, being 100 years old, can house many different layers of energies. A century of hard work, pain, and fatigue, and 100 years of anger and arguments, just swirling and waiting for a living being, an outside energy source from which to draw energy.

Quickly, the three of us said a short prayer and called in Archangel

Michael. We called in angels to blind the room with ethereal light. The results were amazing. In a snap, the room spun, and the air was filled with the sweet scent of flowers! The negative feeling was still mildly present, but was retreating back into its corners, as if pouting.

We gathered everyone back together to do a short EVP session. We needed to inform the group that we felt there was a negative energy present – one, for safety's sake, and two, I wanted to steer the EVP questions in a very specific direction. We just went to all this trouble to push this negative energy or being back away from us and made our intentions clear that we weren't going to engage with such a being. So the last thing I wanted was someone in the group asking, "Is there *anyone* here who wants to talk? Make your presence known." Yeah, let's just invite it in, like some vampire movie. Just like using a Ouija board, if you intend and allow anything and anyone to come on in, they will, and you might not like who or what shows up.

This EVP session was interesting for several reasons. I was very specific in whom I addressed. I only wanted to speak to deceased people, nothing else in this area. I didn't want to invite anything dark into our circle. When you do an EVP session, you usually ask one or two questions and then pause a few seconds, to give the spirits a chance to respond. All I could hear was dogs barking and the resident peacocks yowling during the entire session. We played it back. Where an EVP should have been on the recording, we all heard what sounded like radio interference, like someone was jamming the airwaves. And peacocks. And dogs barking. I thought the whole session had been a waste until one of the investigators pointed out a strange coincidence.

"Do you notice the peacocks and dogs only make noise when we're recording?" she observed. Now they've stopped.

We thought maybe they just went to sleep, but no. We tried recording again, only to have the peacocks start squawking and the dogs start barking again. Every time we hit record, some sort of audible interference would manifest. It was as though something was intentionally disrupting our ability to communicate with the other side.

Could it have been this negative entity somehow causing some sort of static interference? Was whatever negative force that lurked inside these 100-year-old vat rooms angry that we didn't want to interact with it, so was still making its presence known by not allowing us to communicate with anything else?

I had never before encountered anything quite like that. I've had EVP sessions where, when played back, all you hear is static, like some sort of electromagnetic disturbance, but never static timed in specific spots on a recording. It was a little freaky, to say the least. We chalked this up as personal experience since static on a recording doesn't prove the paranormal.

We were anxious to get into the barn loft area as this is frequently a hotbed of paranormal activity. We waited until late in the evening because it had been so warm that day. Even at night our Mel Meters were reading an ambient temperature of 84 degrees. With everyone seated in a large circle in the dimly-lit attic, I tried to get a feel for the room. I wanted to see if any spirits who had worked at the Graber Olive House might still be hanging around. We had several pieces of equipment set up including a couple Mel Meters, a RemPod, and we were running our digital recorders.

I could see a shadowy figure standing on the stairs leading up to the attic. There were strips of leather hanging on the walls and this spirit brushed up against them, making the straps sway in a ghostly breeze. I asked him to please join us in the room, but he seemed so shy. He probably couldn't figure out why a bunch of people were sitting in his attic space.

Another investigator had set up a laser grid just past the stair landing. The green laser lights flickered when the spirit walked past. This helped everyone in the room see where the spirit was moving and provided scientific documentation on what I was seeing. The shy spirit slipped against the wall and joined us, just watching.

Then things got a little weird. The energy in the room began to shift. It became more statically-charged and we could feel a change in pressure. Kimberly was sitting to my right. All of a sudden, everyone in the room began seeing flickers and dashes of lights behind Kimberly. They were different colors – blue, white, green. I have seen color anomalies manifest around Kimberly when she calls in angels, so I asked, "Are you calling them in?"

"No," she said. "It's not me." Something was up.

I began to feel someone standing behind me. I thought it might be the spirit who had come up the stairs, deciding to come in closer. But then all of us started feeling as though someone was standing behind them. Each person described similar feelings; one person said it felt like a blanket was being wrapped protectively around her.

As we sat there, being delighted by each of us having a personal experience, a few of us began to see some odd little spirit entities enter the room. Not one, not three, but more like 20 little spirits, or elementals (I'm not sure what they were), started bouncing into the room. Yes, I said *bouncing*! They appeared to be about 2 feet high and almost as wide, bouncing into the room. They looked like ghostly Oompa Loompas hopping about. *Boing, boing, boing.*

Then they discovered my RemPod. *Beeeeeeeep*! It started alarming like crazy. Now usually when there is a disturbance in electromagnetic fields, the RemPod merely beeps in a short burst. But this time, the RemPod was going off, alarming continuously and loudly. I think a few of these Oompa Loompa spirits must have been fighting over it. I really half expected it to fly across the room – they must have been tugging at it so hard.

One of the docent investigators had just come up the stairs to see what all this noise was. The first thing he noticed was how charged the room felt. With all the beeping and alarms going non-stop, I asked him if he had encountered a RemPod reacting in this manner. He said no. I asked the Oompa Loompas to step away from the device. That didn't work.

This had been beeping constantly for a good two or three minutes by now. I figured I'd better try to debunk it by seeing if it had malfunctioned, but I kind of hated to take their new toy away from them. I picked it up and it still was beeping its alarm. A RemPod is only supposed to beep when something breaks its electromagnetic beam, like a motion detector. Something was physically touching it – something we couldn't see.

With the beeping RemPod in my lap, I fumbled for the off switch. Before I could shut it off, the alarms stopped. I had taken their toy away like a big meanie.

The energy was still crazy in the attic. We had beings standing behind each investigator, the worker spirit was still standing pinned against the wall, probably freaked out by now, and bouncy things were all over the place – oh, and the light anomalies were still flashing. The RemPod didn't go off again, even though I said they could have it. (I guess I spoiled their fun.) There was so much activity all at once.

And then, *poof* – silence. All went quiet. That's usually the case in an investigation; activity reaches a peak and then winds down. They'd had their fun and gone home.

There had been no more activity and by now we all decided to take a much needed water break. The cool night air was a pleasant relief. We sat outside, describing what personal experiences we had just had, all the while munching on s'mores. We were pretty much done with the evening, having had so much activity in such a short time.

There was a short road on the property that leads to a small house. It has huge, ancient trees on either side. Oddly, everyone seemed drawn to this road all at the same time. It was as if the spirits were calling to us one last time. So we all started snapping photos, and then everyone was filled with dozens of bright, glowing orbs.

Now usually I am quick to debunk orbs of light as dust particles or bugs or pretty much anything other than paranormal. But the fact that we all felt the energy, we all felt drawn to this exact spot at this exact time, told me that something was reaching out to us. There were no bugs flying about and there was no dust on the road. It was just a magical end to a very special investigation.

I really felt as though ancient earthly spirits had interacted with us that night. It was like no paranormal encounter I've ever witnessed. I felt truly blessed to have been a part of it.

CHAPTER 27:
Psychic's View of Graber House

The 100 year old Graber House presents some very unique investigating opportunities. While there are some very possessive and shall I say bullying entities there, I particularly like bringing beginning paranormal investigators to this location.

One, it is consistently active. This lends itself to experimenting with feeling the fluctuations in how each area feels. Plus, you are able to gather far more information if you visit a location multiple times. Are you getting the same EMF in the same area each time you visit? The Graber does have a few pieces of equipment that emit high EMF. By knowing this, we can avoid false positive reactions from our RemPod and other EMF detectors. We had to keep our equipment at least 5 feet away from this equipment.

Two, there is a strong, ancient earth energy there. This energy is great to channel. It brings in protection and a sense of peace, like laying in a grassy field.

We conducted not so much a seance but more of a energy circle on our last investigation. We had several folks new to investigations and this was there first time channeling. we wanted a good, protective energy. We needed a positive experience.

Almost immediately we started seeing visions of animals and insects joining in the circle. I'm not talking creepy bugs but these resembled primordial insects from long ago. Soon visions of gem like fire burned in the circle's center. What was so fabulous was not just the psychics in the circle saw all this, not just veteran investigators saw this, we ALL saw these visions.

Then, like a blast of heady perfume, the scent of roses. Did this mean the Virgin Mother had joined us? I don't know, but the energy was so comforting, welcoming and yes, motherly.

Roses were followed by visions of hydrangea and other flowers. Soon sandalwood was all around us. All present could lean in and smell these metaphysical flowers. It was such a wonderful experience.

I looked down. I said, “look at your feet1 They are glowing!” Indeed, everyone’s feet were glowing a pale blue, as if we were all standing on florescent lights.

Everyone’s breath was taken away. it was truly a magical experience. THAT’s what can happen when you bring your own intuitive abilities to a paranormally active location. I hope you have similar,wonderful experiences.

CHAPTER 28:
Star of India

The Star of India in San Diego. Ca

The Star of India is the oldest active sailing ship in the world. Of course, it's VERY haunted!

The Star of India is the world's oldest active sailing ship. She was built at Ramsey Shipyard in the Isle of Man in 1863. Iron ships were relatively new for that time then, with most vessels still being built from timber. She originally bore the name Euterpe, after the Greek muse of music and poetry.

The Euterpe was a full-rigged ship until 1901 when the Alaska Packers Association rigged her down to a barque, her present rig. The advantage of these rigs was that they needed smaller (therefore cheaper) crews than a comparable fully-rigged ship as there were fewer of the labour-intensive square sails, and the rig itself is cheaper. They were still very fast and she could travel from England to India and back in about 100 days. Can you imagine such a journey?

She began her sailing life with two near-disastrous voyages to India. On her first trip, she suffered a collision and a mutiny. On her second trip, a cyclone caught Euterpe in the Bay of Bengal, and with her topmasts cut away, she barely made port. Shortly afterward, the captain died on board and was buried at sea.

After such a hard luck beginning, Euterpe made four more voyages to India as a cargo ship. In 1871 she was sold again and embarked on a quarter-century journey of hauling emigrants to New Zealand, sometimes also touching Australia, California and Chile. She made 21 circumnavigations in this service, some of them lasting up to a year. It was rugged voyaging, with the little iron ship battling through terrific gales, "laboring and rolling in a most distressing manner," according to her log.

Life aboard the Euterpe was difficult for the sailors, cooped up in her 'tween deck, fed a diet of hardtack and salt junk, subject to malnutrition, seasickness, and a host of other ills. It is astonishing that their death rate was so low. They were a tough lot, however, coming from the working classes of England, Ireland and Scotland.

Her name was changed to Star of India in 1906 and, after 60 years of solid performance in 1923, she became the crown jewel of the Maritime Museum of San Diego, where she remains today as a floating piece of history.

We arrived in the afternoon and walked into one of the shops that are located next to the Star of India. The store was crowded with white-suited sailors fresh off their ship and they were eager to do some shopping. I was asking the clerk something when I noticed my body began to sway back and forth, as if rolling on a ship's deck. The store is located within a ship, but it's securely docked. I looked at all the sailors but they weren't swaying, so I figured they had their sea legs.

Now I'm swaying quite a bit, back and forth, and I really felt as though I was going to get knocked off my feet – and the worst part was I couldn't seem to do a thing about it. It was like some kind of out-of-body experience. I knew I was doing it, and I felt like an idiot swaying to and fro, but I was completely unable to control my actions. Finally I grabbed onto the counter and stared wide-eyed at the clerk. Of course she was looking at me funny by now.

I managed to gain control and, as we walked out, Kimberly asked me, "So what was that all about?" She had seen me swaying and she had begun swaying too. Was something trying to communicate with us already? I've never felt like that before. I was feeling like I was slipping out of body non-consensually, unable to control the swaying, even though I was cognizant of the fact that I was doing it. I was helpless to stop myself from tumbling to the floor. It was as if the spirits of the Star of India knew we were coming.

We started our lockdown by being escorted by one of the Star's docents, Dave. Dave walked me through the ship's four decks and located the light panels and alarm system.

"I'm going to be gone, and it's the security guard's fourth day on the job, and during the first three days he didn't have much training," Dave explained. "Make sure when you're ready to leave that you show him where these panels are and make sure he sets the alarm. If you need to leave in a hurry and you're running down the plank, the gate won't be locked, just push through it."

Wait, what? Has he had people running off the ship in a hurry? Dave wouldn't say.

Still wondering if I was going to be running screaming down a poop deck in the middle of the night, we set up our base in the galley and then we headed down below deck. This is another reason I wanted some ambient lights on. Have you tried to navigate a ship's steep stairs with a thin rope for a rail? There was no way I was going to try to do that in complete darkness. Besides, it's a museum too, and I don't want to break anything.

We walked from stem to stern taking still digital shots. I felt drawn to crawling into every nook and cranny of the ship. It just felt as though they were somehow hiding from us. Sure enough, the cameras began malfunctioning. It was as though they were deliberately causing the equipment failure. The camera would turn on, I'd point it in the direction I wanted to photograph, and boom, the camera would lock up.

We had three separate battery drains as well despite the fact that I had made sure all the equipment had fresh batteries before we started. The batteries would go from a full charge to completely drained in a matter of minutes. It's a common occurrence for electronic equipment to drain or malfunction with the presence of paranormal activity. I believe they do this to gather energy from an external source in order to manifest either visually or audibly.

Remember, spirits have no physical body, so in order to be seen or for you to hear a disembodied voice or to create an EVP, they need an external source of energy. I also believe they sometimes just like to screw with us. I believe they mess with equipment when they don't want to be caught on camera. Spirits are very resourceful that way.

So began a cat-and-mouse game of loud bangs in one room and equipment malfunctioning in another, basically leading us from one end of the ship to the other. We decided to regroup on the main deck.

With cameras behaving themselves now, we began taking more still shots. Kimberly's sister Michelle noticed she was getting a lot of orbs immediately in the vicinity of the lifeboat. Over the course of several consecutive shots, dozens of orbs would suddenly appear around the lifeboat, then in the next frame they would slightly disperse, and be gone completely by the fourth shot. There was no dust and very little breeze, so we ruled out dirt contamination. I would take several shots of the same boat with a similar camera to no avail; the orbs only appeared when Michelle took the pictures.

With activity seeming to center close to the lifeboat, we figured this would be a good time and place to conduct an impromptu seance. Each of us grabbed a chair and we sat in a circle. We relaxed our minds and allowed ourselves to slowly focus on the spirits around us. Silently we invited them to join us.

I let my eyes fall slightly out of focus (and at my age that's pretty darn easy) and I gazed up at the elaborate rigging in the masts. Slowly taking shape aloft in the rigging lines and beams were shadowy figures. I gazed a little longer and I could see the haggard, barefoot sailors staring back at us. They looked at our group with what I can only describe as a bit of disdain, they looked thoroughly unimpressed with these ladies trying to demand and order them about.

Just then I distinctly heard the words "bilge rat." It seems the pecking order of sailing ships is certainly Captain, crew, *then* female ghost hunters! This was going to be harder than I thought.

We had been politely asking for general information from the crew but it seemed obvious now that being polite little women was getting us nowhere fast. While we didn't want to stoop to the level of harassing or taunting – that's just rude and unnecessary and we didn't want to anger anyone – we did need to rethink our approach. Who would the captain respect enough to answer their questions?

On a whim, we thought, "Let's tell him we're from the ship's owners and we're here to make a report." We didn't really expect it to work but figured it was worth a shot.

Using our phones, we Googled one of the several owners of the Star and took a guess about the timeline, picking the Shaw, Savill and Alboin Line. The ship was still the Euterpe during the 19th century.

We had one of our male guests, in his best British accent, call out into the night air, "We've come from the Shaw Company and we need to ask you some questions. We need to make a report and we need you to comply, Captain."

I was sitting with my back to the Captain's Quarters. It had a small window and short, cotton curtains with the lamp's light glowing through. Kimberly faced the Captain's Quarters. Her eyes wide, she said, "Kitty, the curtains just moved!" I wheeled around to look. Sure enough, the tiny curtains were parting ever so slightly. There is no way any breeze can get into the Captain's Quarters because of the wooden partitions inside the galley.

We kept talking. "We need to ask you some questions, Captain." The curtains opened a little more. We all saw this happening. We could have squealed with excitement but were trying to play it cool.

We had all of this attention for now, but I knew time was of the essence. We quickly relocated into the galley. We didn't know how long we could keep the Captain's attention. We started running several digital voice recorders and began an EVP session. The Captain's presence could certainly be felt. Kimberly and I swore we could see him sitting on the galley's sofa, glaring at everyone.

It was in these first few minutes of recording we captured a Class A EVP saying, "Why did you lie to me?" We had made direct contact with the Captain of the ship and he understood what we had said. Apparently, the Captain did not appreciate being duped. Would the ship's crew now come forward and speak? We captured a couple more EVPs of conversations between several voices talking excitedly. We were unable to make out what they were saying, but the word "hiding" came through clearly. What were they hiding?

Everyone needed a break. Out came the water, Mountain Dew, grapes, cheese, and Ding Dongs. I had a bet with Kimberly that the Ding Dongs would go first before any healthy snacks. I lost, but the Ding Dongs were awfully good. Recharged with a nice sugar buzz, we headed back into the galley. We thought we'd try a Spirit Box session to see if the ship's crew would come through.

The Spirit Box, also known as Frank's Box after its creator Frank Sumption, is a useful tool to communicate with spirits. It's basically a radio, usually an inexpensive AM/FM radio, that has been modified to continually sweep through all the channels without stopping. Picture turning the dial on a car radio back and forth and you get the idea. What this does is creates a type of white noise, a physical medium for spirits to use as their voice. I find it's better to run the channels in a backwards directional sweep to avoid any stray random words from popping in and creating false evidence.

We sat ourselves around the galley table. It had already been a successful night and I just wanted to take a moment and soak it all in. I looked around the room, the woodwork gleaming with a fresh varnish. We were sitting at the ship's original galley table. This table had been here since the 19th century, before the Civil War. I thought about how many people had sat in these chairs over the decades. I ran my hands across the big, heavy table. It carried its share of scars and history.

This is when you really begin to connect with spirit. Imagine their time in this place. What must they have endured at sea? Did they ever see their loved ones again? Imagine their hardships and the difficult decisions that were made in order to simply survive. They did all this to make a better life. This is what I love about paranormal investigating – when that connection is made. It becomes real.

We were still setting up some equipment when we all heard what sounded like a conversation. Whispered voices were chattering excitedly but I couldn't make out what they were saying. We all heard these voices with our own ears; they were audible spirit voices, not EVPs. Was the ship's crew trying to speak to us? These voices showed up on our recorders but I was unable to make out what they said. The word "hiding" was the only word I could make out of the muffled conversation.

Just then we heard a loud groaning noise. It was like a metal scraping on wood. We stared wide-eyed at each other. It was *loud*. It did it again. We could feel the vibration through our feet. *Groannnnn*. It did it again. We looked around. What *was* that?

One of the guys in our group peeked outside. "Oh, there's a motorcycle outside," he said. Yeah, we heard it. *Vrmmm, vrmmm, groannnnn*. Wide eyes again. "That's no motorcycle!"

Half giddy, half a bit unnerved, we searched for what could be the source of this groaning noise. We searched for the source of the groaning in the back galley. There, the large iron rudder post runs through the ship from the wheel to the ship's rudder. *Groannnn*. The rudder post was rotating. But how? The ship's wheel is securely lashed so it won't move. I figured I better try to debunk this.

"I'll go up and see if I can get it to move," I volunteered. Later, we discovered that immediately after I say this, we captured another Class A EVP: "Damn you!" The Captain was still upset with me.

I go above deck and grab the ship's wheel. It's lashed on both sides and only has a couple inches of play. I pull hard back and forth. I called down to see if that made the rudder post move.

"Anything?" I asked. "Nope. Not a bit," the group responded.

Now you might think this is just movement from the water creating some turbulence and causing either the rudder or the wheel to move. But not only could I not recreate the movement by leaning with all my weight on the wheel, but the movement of the rudder post completely stopped while we were standing in the room watching it. The rudder post didn't budge the entire time we stood there watching it, waiting for it to rotate. We could not debunk this phenomenon. I'm leaving this as unexplained, but I believe the spirits were definitely messing with us.

Back to trying to set up our Spirit Box and digital voice recorders. Sometimes you just have to go with the flow and go where the activity takes you. We figured the Captain was telling us he was still in charge of this vessel. Could he make the rudder post move on command? Would he?

With all our equipment running, I asked, "Captain, was that you? And could you please make the post move again for us?"

Immediately, we heard the now familiar *groaannn*.

"Thank you, Captain," I said. (OK, maybe we squealed a little.) Not only did the groan happen instantly after I asked him to respond, but this was the last time during our investigation the rudder post moved. I continued to ask the Captain questions but received no more responses. I was pretty sure that he was done talking.

"If this is not the Captain, then to whom are we speaking?" A moment later, we heard the name Patrick.

I wanted to know what they were hiding from us. It must have been some contraband smuggled aboard the ship long ago. It was at this point we all noticed we had a strong metallic taste in our mouths.

Gunpowder? Ah-ha! They were smuggling gunpowder. That definitely could be a very dangerous cargo.

So where does one acquire gunpowder in the 19th century? China, I would guess. With recorders and Spirit Box running I asked, "When did you get the gunpowder, and from where?" The answer came on the Spirit Box: China.

It's really amazing to receive this much validation. We had physical movement of the rudder post that rotated on command and then stopped. We had a Class A EVP saying "damn you" – possibly from the Captain. We had voices from the Spirit Box answering our questions intelligently. Remember, the bandwidths are sweeping backwards to eliminate the chance of stray radio voices coming through causing false positives.

It was time to do what we are called to do, and that is to provide guidance in crossing spirits and sending them home. Doing so in a haunted location does not create a vacuum and render a place devoid of paranormal activity. When done with love and the right intentions of helping spirits who wish to return home, I believe it actually creates some sort of portal. Spirits might then communicate that there are folks on this side who wish to learn, talk, and help if possible. I think crossing spirits actually *attracts* other spirits to a location who might also need help.

Kimberly called in her angels and we all focused on filling the room with positive energy. With the Spirit Box and recorders still going, Kimberly said, "Reach out your hands and find that your family is there waiting for you."

"Hi Mom," came from the Spirit Box.

"Your time is served, you are now free to go home," Kimberly responded.

We heard "thank you," and then the most silent bit of silence I've ever heard from the Spirit Box. He was just gone. Cue the tears and goosebumps.

We couldn't top that. It was late and time to call it a night. It was time to find our wayward security guard and help him go through all the decks and set the alarm. Remember, it was only his fourth day on the job. Packed up and ready to hit the showers and a soft bed, we disembarked the lovely Star of India and expected our security guard to be there. But there was nobody. The guard has three ships to patrol so we headed out to the Berkeley.

"John!" we called out. "Where are you?" I decided to call his phone. "Hello, this is John," he answered.

"We're finished up here on the Star and we need you to lock up," I said.

Laughing, he asked, "Did you find any ghosties?"

"Yes, it was incredible," I responded. "We had a ton of activity. This ship is alive with spirits."

There was an awkward silence. Then, in a softer tone, he muttered, "No shit?" He must've started thinking about all those decks, all those dark places on the ship he had to monitor alone, because he then told us to go on ahead, and that he would just leave whatever lights were left on and set the alarm himself.

I wonder if he made it there a full week?

CHAPTER 29:
Psychic's View of the Star of India

I have to admit I was a bit surprised at how much activity we encountered and the amount of evidence we were able to capture during our investigation aboard the Star of India. The spirits were just talking among us nearly the entire time.

In our audio, we had multiple EVPs, and captured the sound of the rudder post moving on command. Those excited whispered conversations everyone heard with our own ears in the galley were caught those on tape as well. For photographic evidence we were able to catch a light anomaly in the room with the rudder post when the post was rotating. Kimberly's sister Michelle captured several frames of orb activity around the lifeboat. And don't forget that wonderful Spirit Box session. Thinking about how we helped a spirit cross over, heard him greet his mother and then thank us still brings us goosebumps.

Once again, how did bringing psychics along enhance this investigation? Let's start with the most basic answer: numbers. You have a large area to investigate and a limited amount of time. You have to find a way to best direct your focus and a psychic can help us find the spots most likely to have paranormal activity. Where are the ghosts?

As always, we start with a full sweep of the location. This is when you walk around, take your still shots, check for ambient or unusually high EMF spikes, locate drafty areas and items that might move or creak that may give off false readings. A lot of times it's good to do this earlier in the day or even on a previous visit. But this perimeter walk is also a great time to see who's with you. Even if you don't consider yourself sensitive, stop and take a moment. Does it feel as if someone is following you? Do you feel drawn to a particular spot and don't know why? Remember I felt compelled to crawl into some of the lowest cargo holds of the ship? This is when my camera began malfunctioning. I believe they were hiding in these lower holds and they caused the camera to stop working. It is common for spirits to manipulate equipment. I think even spirits from the 19th century have somehow learned what cameras and digital recorders do – and if they don't want to be caught on tape, they won't.

A psychic's job is to introduce the group and let the spirits know we are there to learn and to help them if needed. A psychic's job on an investigation is to make that connection with spirit, to help bridge the gap between our world and theirs. Psychics are able to raise their vibration to meet that of the spirit world and hopefully, in turn, they meet us halfway. In the middle. (Hence the term "medium.")

Conducting the short seance gave us a good indication of how the spirits of the Star of India really viewed us. We brought the investigators together by just relaxing our minds and letting ourselves see who is there. Sometimes it's easy to get wrapped up in capturing evidence on film or audio that some of us get pulled away from that connection. You should have seen how excited we were when that rudder post was rotating several inches and we couldn't debunk it. Everybody wants to get that validation, that evidence. By conducting even a short seance it brought everyone back to being on the same page and reestablished that all-important connection with the spirits. Otherwise, we're just tromping around taking haphazard photos of a creepy place and act surprised if we caught some anomaly, not knowing why the anomaly may have shown up in that particular photo.

We figured out the hard way that the hierarchy of the ship – even after death – puts the Captain still in charge. Once we figured that out, it seemed logical to conduct the EVP and Spirit Box session in the main galley, where the Captain directed his ship and crew. It was here that he let us know he's still in charge. I believe it was the Captain that caused the heavy rudder post to rotate. The Captain made it rotate one last time when I asked, and then the post remained still for hours the rest of the night.

What really guided the investigation was us picking up on some smuggled cargo. Just getting that hit psychically gave a direction in our line of questioning. When we started tasting the metallic gunpowder on our tongues, we knew we had them. The Spirit Box – which was running in a backwards sweep to prevent random words from radio stations coming through – gave us answers that were relevant and immediate.

That's when it's imperative that you have that good connection with spirit. We could feel when the Captain was done talking and members of the crew were willing to come forward. A little coaxing toward a reluctant ghost goes a long way.

While the Captain was feeling comfortable where he was in the afterlife, I believe it was one of the crewmembers who accepted guidance in crossing to the light. We agreed he felt so guilt-ridden that we could feel his sense of relief once he confessed about the cargo. It was a pleasure to unite him with family and send him home. This was recorded and validated through the voice from the Spirit Box as he thanked us, and then the Spirit Box went completely – and I mean completely – silent. We no longer even got the short bursts or fragmented words anymore. Just silence. He was gone.

Once again it basically comes down to directing the investigation toward the activity, and by that same token, inviting the spirits closer to the investigators and our equipment. I believe this has given us an edge in consistently getting some sort of evidence at pretty much every location we've visited. Even if you pick up on a spirit's energy, and then they're gone, you've been able to make that contact. When you come back, the spirits may remember your kind intention and be more willing to communicate. I've had that happen. Sometimes you have to visit a place a few times before they're ready.

I want to end this chapter by saying we held this investigation on a Saturday, August 15, 2014. We had one of our best Spirit Box sessions to date. On Sunday, August 16, Frank Sumption, the inventor of the Frank's Box, passed away. It was through his work and research that investigators today can make these strides in the paranormal field. He was a pioneer to the paranormal community and his contributions to ITC research are well respected. Frank was also one of the nicest guys you'd ever meet. RIP Frank Sumption, and thank you.

The Berkeley Ferry in San Diego, Ca

So beautiful and yet such dark entities lurk below deck!

CHAPTER 30:
The Berkeley Ferry

This chapter serves as a reminder that negative and demonic attacks can occur before ever entering a paranormally active location. They seem to always know you're coming….

This investigation started out as a relaxing week of food, friends, and haunts in one of the oldest and most haunted cities in California..San Diego. Earlier in the week we had revisited the sailing ship Star of India, getting some amazing evidence. I remembered this time to bring the captain a small bottle of rum. We were running our recorders when I asked permission to come aboard and wanted to thank the captain for his hospitality with this bottle of rum. Upon review, we heard an EVP " We can have?" Hopefully, that meant he'll share it with the crew. We couldn't visit San Diego without snooping around the many haunted locations in the heart of Old Town. We couldn't resist some impromptu EVP sessions back at the Whaley House and photo shoots in it's 19th century cemetery. The Whaley House was built on the site of the town's original gallows. Mr. Whaley wasn't the superstitious type and got the land for a song. The old cemetery is just a short walk down the road. The street is quaintly illuminated by vintage gas street lights. We sat in the cemetery, enjoying the warm summer night. In the golden light of the gas lamps several of us could make out a shadow figure standing over one of the graves. It was short, about 2 feet high. We watched it for a moment as it was barely 10 feet from us. It moved, stood all of it's height of 2 feet, then it bowed towards the ground. It was a dog! A dog bowing and stretching as if it had been napping over its owner's grave. Then it was gone. It was such an intimate experience just to watch this spirit dog still watching faithfully over it's owner. It was a very touching experience.

While the old Whaley and it's cemetery have endured the test of time, Old Town has grown around it. Now they sit smack in the middle of a very busy tourist lane.Tourists come to shop for souvenirs and trinkets from across the border, check out a haunted cemetery and then go have lunch at one of the dozens of Mexican restaurants across from the Whaley. It's a little weird, but we're used to it. (THE best Mexican food in the world is in Old Town, I'm not kidding).

I like it when a city keeps its historic landmarks right out where everyone can enjoy them. Why hide them away and keep them fenced up and closed, languishing and decaying? I believe in order to preserve historic locations, let folks enjoy them and grow to love them. I think then they will always want to preserve these places for future generations. I love the idea of Old Town San Diego. You walk around these remnants of early California.The Cosmopolitan Hotel, the Casa De Estudillo, the Whaley House, and several museums set up in century old barns and theaters. Then you do a little shopping, and have a bite to eat. It's like going back in time and revisiting history. Walking where early Californians walked and shopped. Maybe a woman in the 19th century bought a new hat in one of these shops to wear to a wedding at the Cosmopolitan. Or perhaps they stabled their horses after a week's journey to attend a party at Casa De Estudillo. These buildings still stand. Visitors still come and enjoy the hospitality. The spirits of these old dwellings continue to greet and interact with visitors to this day. History continues to be written..

I like to remain familiar to haunted sites I've investigated. I remember the sights, the smells, the way a location feels. I notice when it feels welcoming, and when they occasionally feel like the spirits want to be left alone. I believe going back to locations I've investigated in the past can create a kind of connective thread. It becomes a connection to it's history, it's past. Visiting locations, whether you conduct a formal investigation or not, becomes a reminder to the resident spirits of who you are and what kind of person you are. They remember your last visit and will act in kind when you come again. It's sort of like checking up on an old friend.

Old Town San Diego and the Star of India have become welcoming and familiar.

We thought the Berkeley was going to be just another investigation in wonderful old San Diego. We were wrong..

The Berkeley Ferry sits within a stone's throw of the Star of India. Both are an integral part of the San Diego Maritime Museum. Each has it's own tragic history. But we were completely unprepared for the dark and aggressive entities that awaited us aboard the Berkeley.

Built in 1898, the Berkeley operated for 60 years in San Francisco Bay. A California State Historic Landmark, she is considered "irreplaceable".

Although she had only one confirmed death during her active career and a severe collision with the liner SS Columbia that almost sank her, the Berkeley saved countless lives. During the 1906 San Francisco earthquake and subsequent fire, she carried thousands of survivors to safety. The ferry could hold hundreds of survivors as she escorted them to safe harbor. Not knowing of their own families' fate, the captain and crew worked selflessly day and night to help victims escape the flaming city. We wondered if the captain still manned the helm. Was he perhaps trying to continue to rescue souls in eternity?

Such a storied past is sure to leave an impression on the old ship. Such tragedy must have left an energetic imprint in the wood, the metal, the fabric of the Berkeley. Recent reports of paranormal activity include hearing a woman's voice singing, hearing footsteps walking the wooden decks and reports of seeing wet footsteps leaving the restroom long after visiting hours have ended.

But the sequence of events of what should have been a productive and fun night of investigating a new location became a night of fear, paranormal attacks, and survival…

I had been aboard the Berkeley several times. The Berkeley is where I had my first encounter with something that almost completely took over my body, you could call it a partial possession. I was in the gift shop area of the Berkeley. The store was busy, and it was in the middle of the day. Not the expected time or place for spirits to want to channel. I was talking to the clerk about the Star when I found myself rocking back and forth. Slowly at first, I thought maybe the ship was listing a bit. I looked around and the shop was busy with visiting sailors for the upcoming Tall Ships Festival. I noticed no one else was rocking to and fro. Could my sea legs be that weak? I thought perhaps these sailors were more acclimated to a rocking ship. But the Berkeley is securely tied and there would be no way she could start rocking or pitching in the harbor. Why was I the only one rocking back and forth? The more I noticed no one else was rocking, the more I found myself unable to stop my own rocking. It was becoming quite deliberate, from side to side, and I was powerless to control or stop it. It was terrifying and yet fascinating that some outside force was so desperate to communicate that it was attempting to channel through my body. My eyes were wide staring at the clerk as I rolled back and forth and I truly feared I might fall but could do nothing to stop it. By now I was bobbing and weaving in a very exaggerated motion and yet there was nothing I could do. My curiosity about who was doing this and how made me not fight it. Oddly, I didn't feel threatened by what was coming through me. This distraction was becoming quite apparent to the clerk behind the counter and I finally had to grab ahold of the counter to stop the pitching. It stopped. Whatever was trying to communicate, had just been flushing through me. I could hear no communication in words from whatever it was, only energy and emotions. It was the most curious wave of energy. As I walked outside, my friend Kimberly laughed," Well, What was THAT all about?" I had no idea.. No idea at all...

The Berkeley is too lovely a ship for me to hold a grudge. Built like an old Delta paddlewheeler, her open deck design seems inviting. Since that first pitching episode, I've taken many self-guided tours, walked her long decks. I've photographed the neat rows of long wooden benches, original to the ship. They resemble beautiful church pews. This, along with custom stained glass windows, gives the Berkeley an odd, church like appearance.

The Berkeley's engine room is as fascinating as it is unique. Her engines are custom made and like no other. She was the first propeller-driven ferry on the west coast with her custom triple - expansion steam engine. A confusing puzzle of gleaming brass and valves, her engine has been completely restored. While no longer connected to her paddlewheel, docents proudly demonstrate it's running features using compressed air.

The adjacent furnace room, now silent and cold, has a creepy foreboding look about it. The room, once the beating heart of the ship, is almost too quiet. It's like there's a resentment that lingers. It's hard to describe, but you could feel the difference in the energies of the different rooms. One was vibrant and restored and shiny and almost a celebratory feel about it. The pride the docents took when showing off the gleaming pistons and gauges and pipes. Then the sadness, almost despair I felt in the furnace room. Feelings of neglect, and futility and depression. Were these emotional imprints from the laborers who toiled thanklessly in the oppressive heat and dark bowels of the ship's engine room? Just standing in the room made me feel overcome with a deep depression.

The ship felt alive...and angry…

Negative entities can attack well before you set foot on the location. Remember, they know you're coming..

Our investigation began as most successful investigations do, starting with dinner. The group was a mix of seasoned investigators and a few new to the paranormal. We had a good investigation the night before on the Star of India and had confidence in the night's investigation. The spirits of the Star of India were more receptive to us since our last investigation. I think it may have been the bottle of rum I had brought as a gesture of goodwill towards the captain. Besides receiving the EVP saying, " We can have?" in response to the rum, we also captured many EVPs suggesting they were pleased we had women aboard the ship. The spirits were much more receptive to us this investigation. The Berkeley was a much younger ship and we figured we might encounter the usual salty sailors and probably some wayward passengers. I was intrigued by the recent reports of wet footprints. The custodians said they were small, like a child's footprints. Could this be the spirit a child who may have drowned, a child desperately trying to escape the flames of the San Francisco fire in 1906? I wanted to find out.

Not many people investigate the Berkeley, and this was a first time for many of us conducting a full night's investigation aboard her. We were excited to get this paranormal investigation started but something made me feel anxious.it was more than just the excitement of a new location. I wasn't the only one on edge tonight, either. It had been a productive investigation aboard the Star of India, capturing great evidence, but something seemed amiss at dinner. A young couple, "Dan" and "Sarah", had joined us for our investigation the previous night aboard the Star but tonight Sarah was not at dinner. "She has a bad headache, and will try to join us later", Dan said. When Sarah eventually did join us, she was just not right. Sarah seemed preoccupied and nervous. She said she wasn't sure if she wanted to come tonight. We thought it was the headache..It wasn't..

We finished dinner and made our way to the Berkeley. While everyone checked their equipment, I wanted to do a walkthrough with the security guard. He started off by giving us several restrictions of what we could and could not do. This seemed odd, because we had received no such warnings when aboard the Star of India. "No Ouija boards, no pentagrams, and no Runes." He wouldn't say why he only gave warnings about the Berkeley and not the Star. My guess is someone had a bad experience with these and now the guard, and probably the whole staff, knew something dark had been invited aboard…

This certainly put an ominous tone to the start of the investigation. A non specific warning to the start of an already anxious night. We took this as a sign we really needed to discuss how to remain safe during an investigation. It's our responsibility to make sure everyone in the group remains safe physically and knows how to deal with negative attacks. We went over the exercise of how to "own your space" and be cognizant of any outside energetic influence as these can and will take advantage. I visualize a circle of light around me as I walk through a location. Picture the fairy godmothers and their magic wands. You have that power within you. Spirits will be attracted to this confidence and light but I have found it pushes away elementals and lower vibrational energies. We joined hands in a circle and invited our guards, our guides and our loved ones to guide and protect us. We circled together and chanted and toned. The energy in the room lightened and so I shrugged off the weird feelings and we got started with the night's investigation.

Sarah still was a bit uncomfortable in going downstairs to investigate. Kimberly decided to stay with her above deck and work with her and see what might be making Sarah so uncomfortable. Kimberly's years of experience in Reiki and hypnotherapy are invaluable in empowering people to take control of situations.

Sarah had been acting strangely for a couple hours now. She was acting nervous and tense and acted as though we weren't to be trusted somehow. She wanted to be here and investigate one moment and wanted to flee and argue and wouldn't even make eye contact in the next. She needed our help but something was creating fear and distrust. She was already in trouble. Something negative seemed to be already attacking it's first investigator.

The rest of the investigators seemed fine so I gathered the rest of the group and we went downstairs into the cargo hold to do some EVP sessions. I like to do short, or "burst" EVP sessions. I can play back immediately and see what kind of spirits we're talking to and respond to their answers right then. What I have to say is right off the bat we were getting foul language like I've never captured before on EVP. It was taunting and crude and just nasty. At first we thought it might be just "swearing like a sailor" but it wasn't. These were full sentences, too. "Did you fu** her to get here?" "Die" and "Fu** Fu** her today." Yeah, scary stuff. The EVPs had a smugness to them I really didn't like. Hearing these vulgar profanities in a deep voice come through regardless of our line of questioning felt intimidating. Usually, if something like this keeps coming through, I don't engage it, it's trying to bully me and obviously doesn't need my help. However, the fact that clear, class A EVPs were coming through excited some of the other investigators. They wanted to replay it and listen to them again and again. I think this just fed it even more. It was feeding off their excitement. It was getting a response and I believe it was enjoying it. I knew it wasn't a good idea to keep engaging with this dark presence. We decided to move on to another area when one of the investigators came rushing over to me. "You need to check on "Mary". Mary was sitting slumped in the corner, doubling over. I asked. "What is it? "What are you feeling?" "I feel sick, in my stomach, like a sharp pain." she said. "It feels like a veil is coming over me." I placed my hand over her head. I could feel the negative energy. It was mocking, almost laughing. Mary wanted to leave. It was making her nauseous and she couldn't fight it here. I told her to get out of the space and go upstairs. In order to stop this type of negative energetic attack the person needs to feel they are in a safe place. It can be hard to fight off this type of negative entity while you feel you are still under attack.You need to stay calm and focused. Above deck, the nausea subsided and Mary felt better. "Owning her own space", she was able to free herself from this negative or demonic entity.

The night's investigation had barely started and we already had two people being attacked. I had to keep the rest of the group safe from this demonic entity that was taunting us......

I stood in the exact spot Mary stood. Sure enough, something tried to feel its way in. It came at me right in my Solar Plexis, right near the navel. If you want to know what a negative energetic attack feels like, I'll tell you. It felt like a bad touch tickle from a creepy uncle. It knows it is scaring you and it likes it. It's that smugness that makes it just feel dirty and wrong. I wasn't having any part of it's creepy little game. With a wave of my hand, I envisioned my circle of light. I elicited my protective energy and said, "No", and pushed it away.

Kimberly had been working on Sarah all evening. Sending Reiki, pulling the negative energy from her. This is what we do in these situations. We walk the person through the situation and give them the tools to recognize what's going on in their body. An investigator must be empowered in order to handle energetic attacks. Remember when Sarah seemed weird and nervous at dinner? As it turned out, all day long, hours before the investigation, Sarah had been hearing voices. These voices said to stay away from us and not to trust us, that we were evil. The voices in her head called us demons. Sarah became terrified. She'd never heard voices before, and now she was hearing voices telling her frightening things, horrible things. She was too confused and frightened to tell anyone. Sarah finally confided that all evening at dinner when she looked at us, we appeared somehow different. She could no longer recognize our faces. She couldn't see "us". It was as though a frightening mask was covering our faces, like an overlay. She knew she was sitting with friends but voices in her head were making her believe she was seeing monsters. I wish we had we known this beforehand, we could have addressed this negative energy much earlier. Like I said, they attack the vulnerable. Sarah was not as experienced as the other investigators but she was very empathic. She was terrified and didn't know what was happening.

Aboard the Berkeley, Kimberly made Sarah close her eyes because she could no longer trust them. Kimberly kept asking her, "Which feels safer to you, hearing my voice and listening to what I tell you, or what the other voices say?" It was the only way Sarah could push away the voices and with them their terrible hold on her. The frightening images faded. She could recognize us again. This entity was trying to do the same with Mary but I caught it in time. It was a stern lesson to always keep an eye on everyone in your group.

After about only an hour of investigating we needed a break. We needed to regroup. Hey, I get it. Paranormal investigations are exciting. Your adrenaline is rushing. You start hearing voices come through on the recorders and it is amazing. It's easy to get drawn in by the evidence. But this was becoming more dangerous, the voices on the digital recorders more foul and taunting. We needed to stop, if only for a moment. There was definitely something very negative here and it was attacking anyone vulnerable. This was quickly becoming less of an investigation and more of an extreme baptism of knowing how to fight off negative and demonic entities. We weren't going to let this entity frighten us away. We were going to to stand our ground.

This was a ferry ship. It wasn't supposed to be this negative. The Berkeley was instrumental in saving lives. Obviously, someone had once openly invited ANY energy to join them or even worse, intentionally invited negative entities aboard. We thought about the security guard and his warnings. Why had he needed to be so cautious?

We were equally cautious. Whatever was trying to have it's way with us was not getting another foothold. It was becoming a very draining investigation, being attacked energetically and having to defend yourself all night. It even had one of the investigators believing we were monsters, covering her eyes with a frightening veil and whispering horrible threats. We had never experienced anything quite like it. This negative force was trying to push its weight around all night. We witnessed many dark shadows and the vulgar EVPs continued. If this negative entity could cause fear in the living, it was probably bullying the other spirits aboard the Berkeley as well, trapping them there.

Personal boundaries are for your protection but it protects those around you as well. We were dealing with something dark and dangerous. It had the ability to affect people's vision and skew their thoughts, making them question their own judgement. I took this as a challenge. This was a perfect opportunity to show investigators how to take control when a negative entity tries to gain the upper hand. We may be on different planes of existence but we are the ones in body.

We are the ones in control of this situation. We help those in need but we give no quarter, we take no shit! We knew we had to keep going. After a break and some Ding Dongs to raise our sugar levels, we tried again.

I wasn't going to give up. I knew there had to be pockets of positive energy somewhere on the ship. The Berkeley is not a negative location. Her heroic captain and crew saved countless lives. The fact that there was a bullying entity here doesn't mean other spirits weren't here, wanting to be heard. I needed to do a solo EVP session.There had to be souls aboard the ship who wanted to speak and I wanted to give them that opportunity. I wasn't going to let that bullying entity squelch their voices.

I went into a place I knew I could be alone. A quiet place. Just me and spirits who wanted to communicate. Yes, I went into the bathroom. At first, the energy still felt tense. Before I could even start my EVP session, I could feel I was not alone. It felt as though there was someone just outside the ship's window. The windows of the ship's bathroom were open but I was far enough out into the harbor that no voices or sounds could carry from shore or from the rest of the group. They were at the other end of the ship. And yet, I could feel someone just outside the window. But outside was only water. Suddenly, I heard a loud belch! Not an EVP, but a loud, audible belch! A belch only a salty sailor could make! I looked outside, but there was no way anyone could be close enough to make it sound as close as I heard it. Hey, sailors…

I started my solo EVP session. I softly asked if anyone needed help and told them we had the ability to guide them where they needed to be. I listened back.. What I heard on the recorder took my breath away.. "Pleeaase help me…….Kitty!" It was so pleading, and yes, they called me by name! I have only been called by name just a few times but never so pleading. I said a soft, sincere, prayer and asked their loved ones to come and take their hands. I don't have a prepared prayer. I just say what I feel needs to be said, and I say it from the heart. I asked for those wanting help to look and see the bright light and know it leads the way Home. I told them they were not trapped and could go if they wished. There was no more pain, no loss, only friends and loved ones waiting for them. God, it felt good! I needed to know there was work to be done here and there was more to this beautiful ship than what we had encountered here tonight. What a wonderful way to end the evening.

CHAPTER 31: Psychic's View of the Berkeley Ferry

This investigation is a perfect example of incorporating all the tools in your psychic handbag.

Starting with trusting your gut instincts. We knew Sarah wasn't herself. I wish she had the confidence to confide in us early on. But I cannot judge. How confusing and frightening it must have been to hear voices that you cannot trust, and these voices are telling not to trust those around you! Learn to trust and follow your gut. Practice it in everyday life. Should you leave for work early today? Why do you feel you should take the stairs and not the elevator? If salmon is on sale, but you have a feeling your friend who hates fish might drop by, get lasagna.

The same goes with being in a paranormally active location. Does it feel comforting to hear these EVPs, or do you get the feeling it's just playing you? Use your senses to feel the difference in each room. Do you feel drawn to walk to a particular area? Remember Mary? I wanted to see if what Mary was experiencing was an energy in that exact location or was it something that Mary was manifesting. That's why I stood in the exact spot she did, and sure enough, the negative entity came at me.

But I was ready. I had my tools. I acknowledged what I was feeling was not my own, but something coming at me. I visualized my circle of light and pushed it away. I utilized Reiki energy to not only protect mary but to gauge this entity's intentions, so to speak. Reiki is a great barometer. Feeling energy and how it changes and interpreting these changes is at the core of how psychics work.

To protect yourself in haunted locations, you can use whatever works best for you. Many investigators use a prayer of protection before entering. Some wear Catholic medals that have been blessed. For Dumbo it was his magic feather. He believed it would make him fly and indeed Dumbo flew. Visualize what will protect you, be it a circle of light, a sword, or a magic feather. If you believe it will protect you, then you have just empowered it to do so. Yes, your guides and guardians are there to protect you but you are very powerful in "owning your space". That's how it works.

We knew the Berkeley felt uneasy. We knew we had to keep our guard up. We stopped and regrouped at least five times during the investigation. Each time we felt we were feeling too bombarded, we stopped. Don't let the adrenaline lead you into trouble. Take your time and figure out what's going on here. Documenting the fact that several investigators were subjected to energetic attacks is just as important as any EVP. When someone is feeling sudden nausea or pain, it makes me prod a bit further. If the symptoms subside when they leave the area as quickly as they manifested, it may be an energetic attack and not a fish taco. On several investigations I have felt sudden emotions or queasiness, only to have it dissipate as soon as I leave the area.

Keeping people safe is paramount. That's why we kept Sarah and Mary secluded so we could focus on them away from what was attacking them. But I knew we needed to stand our ground. What would have happened if we had bolted out like a Scooby Doo cartoon? It would have become that much smarter and stronger. Maybe that's what the last person who encountered this had done. Maybe they thought it would be "cool" to sneak a Ouija board in and were naive to just give an open invitation. The Ouija board is a neutral tool. The user empowers it with their "open invitation". Your thoughts and intentions are powerful tools.

We empowered ourselves and stood our ground. And yet we did not taunt or provoke.

But my favorite part of the investigation was my solo EVP session. It was just me and the spirits. No negativity, no drama. Just communication and healing at it's best. A pure connection to spirit and scientific documentation to prove it. I want that for every investigation.

CHAPTER 32:
Lessons and Gratitude

I have had the deep honor and privilege of being able to work with some of the most prominent leaders in the paranormal community. Without their passion, drive, and consummate professionalism, I doubt the field would be where it is today. Working and training with the best in the field really is the kind of training that will last a lifetime.

I wanted to take this moment to express my gratitude for all of these professionals who took the time to share their expertise and knowledge. Here are a few highlights of some of these investigations.

Jason Hawes and Grant Wilson – "Ghost Hunters"

Surprisingly these guys are very quiet during an investigation. They are really the type who pick a spot and camp out. They wait and get a feel for a particular area.

We were in the Boiler Room in the bowels of the Queen Mary. This place is a dark, rusty iron room with 30-foot ceilings. It's a very foreboding place. Joining us was Bill Chapell of Digital Dowsing. This room always has a feeling that something is just about to happen, like someone is just behind a door waiting to jump out at you. The room is all metal, gutted out with walkways added for safety. Suddenly, we heard a huge crash. It sounded like someone threw a heavy glass bottle with force. It definitely had force behind it. Grant bolted over to the area completely expecting to see smashed glass everywhere. There was no glass to be found.

Jason and Grant have been in the industry well over 20 years. They know that what you see on their shows is a heavily-edited version of what took several days to film and hours of audio, most of which revealed little paranormal activity. It is these two who best instill the lesson of patience during an investigation.

If you head out to a haunted location and every visit yields full body apparitions, you should be suspect. Embracing the times of non-activity gives you more validation of the true evidence even more.

Interesting Factoids

Jason has a tattoo of a bishop's chess piece on his wrist, in honor of his daughter, his princess.

Grant is an accomplished pianist. He also loves elves and fantasy realms. He has even developed his own written Elfin language and sports tattoos with his own Elfin words.

Mark and Debby Constantino – EVP specialists

Mark and Debby were two of the most talented people to investigate with and I loved every chance I got to join them on a location. They came in using the minimum equipment and got fantastic results. Re-read that a few times. You don't have to come in with a truckload of gadgets and 20 static camcorders to gather credible evidence. They captured Class A EVPs while using just a few digital recorders and flashlights.

What they taught me was it's a good idea to run several digital recorders during each EVP session. They used Sony models that have variable sensitivity settings. They would do short burst EVP sessions with three or four recorders, each set at a different sensitivity setting. After a few questions, Debby would review each one. If a recorder got an EVP at a particular setting, then they'd switch them all to that setting. They also felt that holding a recorder gives you a closer bond and yields better results. By doing these short burst sessions, you can hear if you made a connection. It helps lead the line of questioning.

For example, say you asked, "Can we bring you anything from the physical world?" and you get an answer of "tobacco," you can tell them you heard them and ask what else you can bring. It's an immediate way of letting you know you can hear them. I think this is a more productive method than taking all the audio back and reviewing all of it days later. Sure, you might find a few odd EVPs and add them to the results, but it's that instant connection that's made that really counts.

Interesting Factoids

Mark and Debby were good friends of mine and a real asset to have on an investigation. They were always generous with their time and vast knowledge of the paranormal field. Debby always had a knack for getting EVPs in a room when no one else could. I could sit right next to Debby with recorders running and she would capture fabulous EVPs while I might get zilch. It was amazing and frustrating at the same time! never knew how she did that. They will truly be missed..

Chris Fleming – psychic, speaker, investigator

Chris has the most wonderful energy. He works with angels and spirit guides. Chris just has a presence about him that I sum up in one word: devotion. Devotion to his faith, devotion to working with spirits and seeking answers. Devotion to teaching others an approach to investigations with a real desire to connect with the needs and emotions of the dead.

One time we were in a cargo hold area of the Queen Mary. This area was once used to hold and transport German POWs. We thought we would use some trigger objects. We had two young men wearing vintage German uniforms and they brought with them a Nazi flag. We had an Ovilus running. The Ovilus continually repeated the word "history." We were lucky to have a man in our group who spoke German.

We started an EVP session, having him ask questions in German while Chris asked the questions in English. We picked up on some soldiers who were still there. They were lost and confused. In German, the man told them, "You don't need to stay, the war is over." The response was heartbreaking. "They left me. They left me." I get the chills every time I re-live that, it's just so sad.

Chris and I also investigated Dave Oman's house in Beverly Hills. This is the house featured years ago on "Ghost Hunters" as well as "Ghost Adventures." It had been six years since Chris had been there and he was more than excited to see what the house on Cielo Drive had in store. It was at Dave Oman's house that Chris introduced Jason Hawes and Grant Wilson to the K2 meter and how it can be used for simple yes/no communication.

In my first investigation with Chris at Dave Oman's house, we caught an EVP of a young male saying "Oh my *God.*" Chris got this smirk on his face. He says he thinks he knows why the spirit said that. "Tell me what you see," Chris asks him. We play it back and you hear the spirit say, "I see angels." Chris starts crying and I realize what has just happened. Chris works closely with his guides and with angels. The spirits were actually seeing Chris's angels! The experience was absolutely amazing.

As mentioned in the chapter on Oman house, it was on one of my many investigations using the Spirit Box with Chris that we captured what we believed was the voice of Sharon Tate. Chris and I decided to end the evening with a prayer for those who might need help. These can be any sincere prayer. Chris hit record and asked for angels and loved ones to please reach out and take the hands of those who need them, to bring them up and take them home. If you say these right, you might be tearing up a bit. We played back the recording and, interspersed throughout the recording, were EVPs of "thank you." It was very touching to help them in their journey home.

Chris brings to an investigation the kind of compassion that is unmatched. He is one of my favorite people to have with me on an investigation.

Interesting Factoids

Chris Fleming's dad was the famous Chicago Blackhawks hockey player Reggie Fleming

Zak Bagans and Aaron Goodwin – "Ghost Adventures"

The buff, bouncy boys of "Ghost Adventures," these guys would put the fun in funeral. They bring with them an enthusiasm that's only matched by their thirst for answers. Their big energy works well in locations where spirits may be a bit shy. Zak and the boys may be boisterous but they gear their investigations in relation to the location.

For instance, if they are in a sanatorium or a place where illness or disease took many lives, they tone it down and come in from an apologetic angle. If they are in a prison, you'll see them establishing personal boundaries and you'll notice a difference in their posture. Remember where you are and be compassionate about who died there.

I was fortunate enough to investigate the Queen Mary with Zak and Aaron. This was their first time aboard the ship. I did capture several EVPs and all of us had some personal encounters during that investigation. I look forward to another investigation with the GAC.

I can associate with their love of history. Like them, I prefer to investigate historic locations as opposed to private residences. You get to research the history and background of the place. This gives you an idea of who or what will be there so you have a direction for your investigation. You know where residual hauntings may be and where you might find intelligent spirits waiting to communicate. You add pages to the history of these locations every time you make a connection.

Interesting Factoids

Zak adopted his dog, Gracie, from a shelter in Las Vegas, and continues to do charity work to promote pet adoption. And as brave as he he is fighting demonic entities on an investigation, Zak seems deathly afraid of clowns! Well, to be honest, who isn't? They're creepy.

Aaron came to the group with a background in filmmaking. He has a clothing line known as Big Steppin' and much of the profit goes to charity.

Dave Schrader – Darkness Radio

I've investigated several times over the years with Dave. He would host Darkness events at some of the country's top haunted locations. Dave could always gather a top notch roster of speakers and made all of his events run smoothly and a joy to attend. On larger events, Dave would hold charity auctions with all proceeds going to Shriner's hospitals.

I think with Dave's hard work, many folks got a chance to investigate locations they might not otherwise have been able to access.

Interesting Factoids

While we were conducting our investigations, Dave was known to lurk about in a custom-made adult size Jumpin' Jammer pajama set – complete with footsies.

Adam Blai – demonologist

Adam Blai holds a MS from Penn State in Adult Clinical Psychology. He has done psychological work in community and forensic settings. Adam is a peritus of religious demonology and exorcism for the Roman Catholic Church and trains priests in exorcism on a national level. Adam also works as a masters level therapist for the Pennsylvania Department of Corrections as a Psychological Services Specialist.

It was Adam who instilled in me the strong sense of morality in doing paranormal investigations. You have to come in with an intention of healing and assisting where you can. To come in demanding responses from spirits who died traumatic deaths is cruel. Think about it this way: would you want somebody treating your mother this way?

Adam has a unique approach to investigations. He doesn't really engage with spirits for religious reasons. He merely asks if they need help or need prayer. That's all. He will say a prayer for the dead to help cross over. He does not participate in paranormal investigations with the intent to collect evidence. His purpose is to serve those in need.

We were on an investigation on the Queen Mary in the cargo hold. This area held POWs and injured soldiers. We had made a connection with a nurse spirit who said she had stayed to care for the soldiers. We were getting great evidence, including capturing an apparition on film. Unbeknownst to us, Adam had made his way down the stairs and taken a seat. Quietly, he asked his two questions. Saying his prayer, he was able to help the nurse ascend. Suddenly, all our equipment went dark. Everything just stopped.

"Oh, hate to be a buzzkill, I sent her on her way," he said. I had no idea it could be that immediate. It was powerful.

Interesting Factoids

Although he takes his work very seriously and works closely with the Roman Catholic Church, Adam does have a fun side. On one investigation, Dave Schrader's mom came dressed as what might be described as a fairy costume – complete with glitter wand. Adam, being a man of the cloth, blessed the liquid in the glitter wand, thus making it holy water.

Adam then decided he should lead the paranormal investigation (for everyone's protection, of course). Adam Blai – tall, bald, imposing demonologist, leading a paranormal investigation with a blue, glitter fairy wand.

Peter James

I unfortunately never had the pleasure of working with the great psychic Peter James. I still feel indebted to him for his contributions to the field. Peter James appeared many times on the television show "Sightings." It was well known that his favorite haunted location was the RMS Queen Mary in Long Beach, California. His manner of investigating and connecting with spirits was akin to your favorite uncle coming for a visit. He was very vocal and authoritative. He said he was able to see spirits as clearly as a living person. His calm but authoritative manner actually seemed to relax spirits and he was known for having lengthy audible conversations with spirits. He and the spirit Jackie on the Queen Mary were recorded having many such conversations.

His style and and willingness to converse with spirits like they are family was what made him one of the most successful and beloved psychics and paranormal investigators the field has known. He is missed.

Interesting Factoids

Peter loved to sing "London Bridge is Falling Down" to the spirit Jackie aboard the Queen Mary. She would often sing back to him. These audible songs were captured several times on tape and aired on "Sightings."

CHAPTER 33:
Celebrity Outtakes

Jason Hawes

I won a bidding war for the shirt Jason Hawes was wearing during a charity auction and made him strip in front of the crowd to give me the shirt. I really wanted to bid on one of his used Roto Rooter uniforms until he said, "Why would anyone want that? It's got poop on it!" He had a point.

Grant Wilson

Grant bid 2,000 bucks on a signed script from the cast of "Lord of the Rings" in a charity auction. Jason Hawes yelled out, "Hope they sign us for another season!" That was eight seasons ago.

Chris Fleming

While investigating Dave Oman's in Benedict Canyon, Chris found himself carrying Dave's 20-pound Toyger house cat throughout most of the investigation because the cat kept meowing during EVP sessions.

Adam Blai

It's worth repeating that Adam blessed the fluid in a glitter fairy wand, thus rendering it holy water – and then proceeded to carry the blue glitter fairy wand with him as a form of protection against evil. Hey, who am I to judge?

Dave Schrader

Dave reportedly wanted to name a son Garth, so his son would be

Garth Schrader.

Brett Griffith

Brett stops at every garage sale he sees, looking for used tripods. It seems he's a bit hard on equipment and breaks tripods and K2s with some regularity.

Zak Bagans and Aaron Goodwin

A group of 40 investigators (mostly girls, surprise) were being split up into groups to work with Aaron and Zak. Zak calls out across the room of girls, "Hey, Aaron, can you get me a digital recorder or a Mel Meter? I forgot, I don't have any equipment." Aaron, unable to resist, answers, "That's what *she* said!"

Bibliography

Belanger, Jeff: *Communicating With the Dead: Reach Beyond the Grave*, New Jersey: Career Press 2005

Bodine, Echo: *The Gift: Understand and Develop Your Psychic Abilities*, California: New World 2003

Breverton, Terry: Breverton's *Phantasmagoria: A Compendium of Monsters, Myths, and Legends*, Connecticut: Lyons Press 2011

Dyer, Dr. Wayne: *The Power of Intention:Learning to Co-create Your World Your Way*, California: Hay House 2004

Emoto, Masaru: *Love Thyself: The Message from Water 3*, California: Hay House 2004

His Holiness the Dalai Lama with Cutler, Howard C. M.D.: *The Art of Happiness: A Handbook for Living*, New York: Penguin Putnam 1998

Kiester, Douglas: *Stories in Stone, A Field Guide to Cemetery Symbolism and Iconography*, Utah: Gibbs Smith 2004

Pickering, David: *Dictionary of Superstitions*, London: Wellington House 1995

Tesla, Nikola and Childress, David H: *The Fantastic Inventions of Nikola Tesla*, Illinois: Adventures Unlimited 1993

Internet Resources

Brown Lady of Raynham: Selected Extract from "Ghosts of East Anglia" http://home.worldonline.co.za/~townshend/dorothywalpole.htm (accessed 2013)

Cicap: Houdini and Conan Doyle: The Story of a Strange Friendship http://www.cicap.org/new/articolo.php?id=101009 (accessed 2014)

Houdini, His Life and Art. http://www.thegreatharryhoudini.com (accessed 2013)

Houdini Museum. http://houdini.net/museum/ (accessed 2013)

Journal of Neurophysiology: Inaudible High-Frequency Sounds Affect Brain Activity: Hypersonic Effect http://jn.physiology.org/content/83/6/3548 (accessed 2014)

RMS Queen Mary: http://www.ocean-liners.com/ships/qm.asp (accessed 2013)

National Institute of Environmental Health Services: Electric & Magnetic Fields http://www.niehs.nih.gov/health/topics/agents/emf/ (accessed 2014)

Naval History; HMS Curacoa Casualty List http://www.naval-history.net/xDKCas1942-10OCT.htm#curacoalost (accessed 2013)

San Diego History: Collections of the Jerry Macmullen Library at the San Diego Maritime Museum, "Euterpe", Diaries, Letters & Logs of the "Star of India" as a British Emigrant Ship" http://www.sandiegohistory.org/journal/92winter/macmullen.htm (accessed 2014)

Smithsonian.com: The Fox Sisters and the Rap on Spiritualism http://www.smithsonianmag.com/history/the-fox-sisters-and-the-rap-on-spiritualism (accessed 2013)

Stanley Hotel http://www.stanleyhotel.com (accessed 2013)

The Guardian: Who Killed Superman?
http://www.theguardian.com/film/2006/nov/18/features.weekend1 (accessed 2014)

Web.Archive: The Ghost in the Machine, Vol 62,No 851 Published in the Journal of the Society for psychical Research
http://web.archive.org/web/20060621050809/http://www.ghostexperiment.co.uk/ghost-in-machine.pdf (accessed 2013)

World Health Org: Electromagnetic Fields (EMF): Definitions and Sources
http://www.who.int/peh-emf/about/WhatisEMF/en/
(accessed 2014)

WW2Today: HMS Curacoa http://www.naval-history.net/xGM-Chrono-06CL-Curacoa.htm (accessed 2014)

About the Author

Kitty Janusz grew up in a paranormally active house where ghostly events would occur almost daily. Kitty and her family came to consider this spirit as just another member of the household. Still living in the quaint old city of Whittier, Ca, Kitty began researching historic locations and investigating the paranormal around 1990. Her favorite local haunt is the RMS Queen Mary.

Her paranormal investigations led her to focus on EVPs, or Electronic Voice Phenomena. Kitty found she was able to capture evidence of the paranormal through intelligent audio responses more predictably than photographic evidence. "Orbs can be so subjective."

During her years of investigating historic haunted locations, Kitty has personally investigated with some of the top names in the paranormal field. Some notables include Jason Hawes and Grant Wilson of Ghost Hunters, Zak Bagans and Aaron Goodwin of Ghost Adventures, EVP specialists Mark and Debby Constantino, psychic Chris Fleming, Patrick Burns, author Jeff Belanger, demonologist Adam Blai, psychic Chip Coffey, inventor and founder of Digital Dowsing Bill Chappell as well as Darkness Radio's Dave Schrader.

Kitty's years as an investigator and capturing spirit voices asking for help led her on a path to enhance her own psychic abilities. She needed a way to communicate with the dead directly.

In this book, Kitty shares her journey on expanding her awareness of the spirit world which surrounds us, and how each of us can become

educated, responsible researchers of the paranormal.
Kitty appears on the weekly radio show, Into the Light Paranormal, talking to psychics, authors, UFO hunters and interesting folks who live life “On the fringe”. Kitty also regularly hosts paranormal events at historic locations, giving insight into the history of each location and how to use one’s own intuitive abilities to connect with spirit.

Kitty lives in Whittier, Ca, with her husband, Joseph, two cats and her Bull Terrier, Basset Hound mix, Alice. Kitty has always been a strong advocate of animal rescue groups and removing misconceptions surrounding “Bully” breeds.

Kitty continues to give lectures, assist in metaphysical classes and still finds time to go on the occasional ghost hunt. Even if it’s spending the afternoon at an old cemetery taking photographs, it’s a good day….

Made in the USA
Las Vegas, NV
01 March 2021